Human-Centered AI in Engineering Design
by
ant

Table of Contents

- **Interdisciplinary Collaboration**
- **Call to Action: Shaping an Ethical AI Future**

Chapter 1: The Convergence of Humans and Machines

The very essence of engineering, at its core, has always been about solving problems and improving human lives. For centuries, this meant building bridges that connect communities, developing machines that ease labor, and creating systems that enhance our safety and well-being. However, as technology has become more complex and pervasive, the *how* and *for whom* of engineering have evolved significantly. This evolution has led us to the crucial concept of human-centered engineering, a philosophy that places the human user—their needs, values, capabilities, and limitations—at the absolute forefront of every design and development decision. It's a paradigm shift from technology-driven innovation to human-driven innovation, ensuring that the tools we create genuinely serve to empower us, rather than complicate our lives or, worse, cause unintended harm.

The journey toward a truly human-centered approach wasn't a sudden leap but a gradual realization, spurred by the increasing complexity and impact of the technologies we engineer. Early engineering focused on functionality and efficiency, often with little regard for the end-user's experience. A bridge was deemed successful if it stood firm and carried traffic, regardless of whether its design was aesthetically pleasing or intuitive to navigate. Similarly, early industrial machinery was lauded for its output, even if it subjected operators to hazardous conditions or immense physical strain. The primary goal was utility, a direct translation of a practical need into a physical or mechanical solution.

However, as computing power began to grow and systems became more intricate, particularly with the advent of personal computers and early software applications, the limitations of a purely functionalist approach became apparent. The mid-to-late 20th century saw the nascent stages of what would become user-centered design, primarily within the realm of human-computer interaction (HCI). This era was marked by a growing awareness that the way humans interacted with machines mattered immensely. Frustration with complex command-line interfaces and clunky software led to the realization that a machine's effectiveness was inextricably linked to its usability. The development of graphical user interfaces (GUIs), pioneered by entities like Xerox PARC and popularized by Apple and Microsoft, was a monumental step. These interfaces, with their icons, windows, and mouse-driven interactions, were designed with the human user in mind, aiming to make computing more intuitive and accessible. This wasn't just about making software "prettier"; it was about fundamentally rethinking how people could interact with complex digital systems by leveraging their natural cognitive abilities and existing knowledge.

These early HCI principles laid the groundwork for a more profound understanding: technology should adapt to humans, not the other way around. This marked a critical divergence from the engineering mindset that expected humans to adapt to the constraints of the technology. The focus began to shift from simply asking "Can it be built?" to asking "Should it be built this way, and how will it impact the people who use it?" This shift was fueled by research into cognitive psychology, ergonomics, and design theory, all contributing to a richer understanding of human perception, attention, memory, and motor skills.

The evolution continued with the rise of the internet and the explosion of digital services. Web design, and later mobile app development, further cemented the importance of user experience (UX). Designers and engineers began to understand that user satisfaction, engagement, and even loyalty were directly tied to the quality of the user's interaction. Concepts like information architecture, usability testing, and user journey mapping became standard practice. The goal was no longer just to provide a service or a function, but to create a seamless, enjoyable, and effective experience for the user. This was human-centered thinking in action, extending beyond the mere interface to encompass the entire lifecycle of user interaction with a digital product or service.

However, as we stand at the cusp of a new technological revolution driven by Artificial Intelligence (AI), the scope and urgency of human-centered engineering have broadened immensely. AI systems are not just tools; they are increasingly autonomous agents, capable of learning, adapting, and making decisions that have profound real-world consequences. They are being integrated into every facet of our lives, from healthcare and transportation to finance and education. This pervasive integration means that the potential for both immense benefit and significant harm is amplified. A poorly designed AI system, or one that operates without a deep understanding of human values and context, can perpetuate bias, erode privacy, diminish autonomy, and create new forms of societal inequality.

Therefore, human-centered engineering in the age of AI is not merely an enhancement of earlier user-interface principles; it is an imperative. It demands a more holistic perspective that considers not just the immediate user interface but also the underlying data, algorithms, ethical implications, and long-term societal impact. The core philosophy remains the same: technology must serve human needs, values, and capabilities. However, the complexity of AI necessitates a more rigorous and deliberate application of this philosophy.

This means that when we design an AI-powered system, we must begin with a deep empathy for the intended users. Who are they? What are their goals, their frustrations, their cultural contexts? What are their inherent capabilities, and where do their limitations lie? How will this AI system interact with them on a cognitive, emotional, and social level? Critically, how can we ensure that the AI system amplifies human capabilities, fosters autonomy, and upholds human dignity, rather than diminishing them?

For instance, consider the development of an AI system designed to assist judges in sentencing decisions. A purely function-driven approach might focus solely on analyzing historical case data to predict recidivism rates. However, a human-centered approach would insist on understanding the judge's role, the nuances of legal judgment, the principles of justice, and the potential for algorithmic bias to disproportionately affect certain demographic groups. It would prioritize transparency, allowing the judge to understand *why* the AI made a particular recommendation and whether it aligns with legal precedent and ethical considerations. It would aim to augment the judge's decision-making process, providing valuable insights, rather than replacing their critical judgment. The system would be designed to empower the judge with better information, ensuring that justice remains a human endeavor, informed by AI but ultimately guided by human values.

Similarly, in the realm of personalized education, an AI tutor should not merely adapt the difficulty of problems. A human-centered approach would consider how the AI can foster a student's intrinsic motivation, build their confidence, and support their development of critical thinking skills. It would involve understanding how students learn best, the importance of social interaction in learning, and the potential for AI to create learning environments that are supportive and engaging, not isolating or overly prescriptive. The AI would be a partner in learning, designed to understand the student's individual learning journey and provide tailored support, much like an exceptionally skilled human tutor, but with the added benefit of data-driven insights and tireless availability.

This user-first approach acts as a guiding compass for all subsequent design and development decisions. It informs every stage of the engineering process, from initial concept generation and requirements gathering to data collection, algorithm selection, interface design, testing, and deployment. It ensures that the focus remains steadfastly on creating technology that is not only functional and efficient but also ethical, equitable, and empowering.

The historical context, therefore, is not just a story of technological advancement, but a narrative of learning and adaptation. From the rudimentary user interfaces of the early computing era to the sophisticated AI systems of today, the underlying principle of designing for humans has remained constant, yet its complexity and importance have grown exponentially. Early milestones in usability and interface design taught us that user experience is paramount. The current era of AI challenges us to extend this thinking further, integrating ethical considerations, societal impact, and the very notion of human agency into the fabric of our engineering practices.

This user-centric philosophy is the bedrock upon which

we will build our exploration of AI engineering throughout this book. It provides the essential framework for navigating the intricate landscape where human needs and artificial intelligence converge, ensuring that the future we engineer is one that benefits all of humanity. It's a commitment to building technology that understands us, respects us, and ultimately, serves us, empowering us to achieve more and live better lives. The goal is to ensure that as machines become more intelligent, humanity remains at the center of the innovation equation, guiding the direction and purpose of these powerful new tools. This is the promise and the challenge of human-centered engineering in the 21st century.

The integration of Artificial Intelligence (AI) into the engineering landscape marks a paradigm shift, ushering in an era where complex problems are tackled with unprecedented sophistication and efficiency. Far from being a nebulous concept confined to science fiction, AI in engineering translates into tangible tools and methodologies that augment human capabilities, streamline design processes, and unlock novel solutions. At its heart, AI within engineering is about empowering engineers with intelligent systems that can analyze vast datasets, identify patterns, predict outcomes, and even generate design options. This isn't about replacing the engineer, but rather equipping them with an incredibly powerful co-pilot, capable of handling tasks that are computationally intensive, repetitive, or require processing information at a scale beyond human capacity. The focus is on enhancing engineering outcomes, leading to systems that are not only more performant but also more adaptive and responsive to their environments and users.

One of the most immediate and impactful applications of AI in engineering is its role in generative design. Traditionally, engineers would manually iterate through countless design possibilities, often guided by experience, simulation, and intuition. This process could be lengthy and resource-intensive. Generative design tools, powered by AI algorithms, flip this model on its head. Instead of the engineer defining every parameter, they set the goals, constraints, and performance requirements – such as material properties, load-bearing capacity, manufacturing methods, and cost limitations. The AI then explores a vast design space, autonomously generating hundreds, thousands, or even millions of design options that meet these criteria. This allows engineers to discover novel, often counter-intuitive, yet highly optimized designs that might never have been conceived through traditional methods. For instance, in the aerospace industry, generative design has been used to create lighter, stronger aircraft components by optimizing material distribution based on stress simulations. This not only reduces the weight of the aircraft, leading to fuel efficiency, but also enhances structural integrity. Similarly, in the automotive sector, AI-driven generative design is employed to create lighter vehicle chassis and engine parts, contributing to better fuel economy and performance, while also considering manufacturability and assembly constraints. The engineer's role then shifts from manual iteration to strategic guidance, selection, and refinement of the AI-generated options, leveraging their domain expertise to choose the most suitable design for production and real-world application.

Beyond design generation, AI is revolutionizing simulation and analysis within engineering. Complex systems, from intricate mechanical assemblies to large-scale infrastructure projects, require extensive simulation to predict their behavior under various conditions. AI can significantly accelerate and enhance these simulations. Machine learning models, trained on historical simulation data or real-world sensor readings, can create surrogate models that approximate the behavior of complex physical systems with remarkable speed. This means engineers can run thousands or millions of simulations in the time it would have previously taken to run a handful, allowing for more thorough exploration of design spaces and a deeper understanding of potential failure modes. For example, in the development of new materials, AI can predict material properties and performance under different environmental stresses much faster than traditional empirical testing or detailed computational fluid dynamics (CFD) or finite element analysis (FEA). This predictive capability allows for rapid screening of potential material candidates, accelerating the discovery and implementation of advanced materials. Furthermore, AI can be used to optimize simulation parameters themselves, ensuring that the most critical aspects of a system's behavior are accurately captured while reducing computational overhead. This is particularly valuable in fields like computational fluid dynamics, where optimizing mesh generation and solver settings can drastically reduce computation time without sacrificing accuracy, enabling engineers to explore more design variations.

Optimization problems are another fertile ground for AI in engineering. Many engineering challenges involve finding the best possible solution from a vast array of possibilities, subject to numerous constraints. This could range from optimizing the layout of components on a circuit board to scheduling complex manufacturing processes, or even optimizing the energy consumption of a smart grid. Traditional optimization algorithms can struggle with the sheer scale and complexity of these problems. AI, particularly through techniques like reinforcement learning and genetic algorithms, can effectively navigate these complex, multi-dimensional search spaces. Reinforcement learning agents can learn to make sequences of decisions that lead to optimal outcomes, adapting to changing conditions in real-time. For instance, in logistics and supply chain management, AI can optimize routing for delivery fleets, dynamically adjusting routes based on real-time traffic conditions, weather, and delivery priorities to minimize travel time and fuel consumption. In manufacturing, AI can optimize production schedules, machinery utilization, and resource allocation to maximize throughput and minimize downtime, responding intelligently to unexpected disruptions like equipment failures or material shortages. This ability to learn and adapt makes AI-powered optimization solutions far more robust and effective than static, pre-programmed approaches, especially in dynamic and unpredictable environments.

The concept of "smart" or "adaptive" systems is increasingly becoming a reality thanks to AI. Engineering is moving beyond static, predictable machines to systems that can perceive their environment, learn from their experiences, and adjust their behavior accordingly. This is crucial for applications in robotics, autonomous vehicles, and intelligent infrastructure. AI enables robots to not only perform pre-programmed tasks but also to navigate complex environments, recognize objects, interact with humans safely, and learn new skills through observation or trial and error. In the realm of autonomous vehicles, AI is the core technology, processing data from sensors like cameras, lidar, and radar to perceive the surroundings, predict the behavior of other road users, and make driving decisions in real-time. This requires sophisticated AI models for object detection, tracking, path planning, and control. Similarly, AI is being integrated into smart grids to optimize energy distribution, predict demand fluctuations, and manage renewable energy sources more effectively, leading to greater efficiency and reliability. Infrastructure itself can become smarter; AI can analyze sensor data from bridges and buildings to predict structural integrity issues before they become critical, enabling proactive maintenance and preventing failures. This transition to adaptive systems fundamentally changes how we engineer, requiring a deep understanding of AI's learning and decision-making capabilities.

Furthermore, AI is playing a pivotal role in predictive maintenance, a proactive approach that significantly reduces downtime and maintenance costs. Instead of relying on fixed maintenance schedules or waiting for equipment to fail, AI analyzes data from sensors embedded in machinery – such as vibration patterns, temperature readings, and acoustic emissions – to identify subtle anomalies that indicate potential future failures. Machine learning models can be trained to recognize these early warning signs, allowing maintenance teams to intervene *before* a breakdown occurs. This shift from reactive or scheduled maintenance to predictive maintenance is a major advancement. For example, in the aviation industry, AI-powered predictive maintenance systems can monitor the health of aircraft engines, alerting airlines to potential issues weeks or months in advance. This allows for scheduled maintenance during planned downtime, avoiding costly flight cancellations and ensuring passenger safety. In manufacturing, predictive maintenance can optimize the upkeep of assembly lines, minimizing unexpected stoppages that can disrupt production and incur significant financial losses. The insights generated by AI also help engineers understand the root causes of component wear and failure, feeding back into the design process to create more durable and reliable products in the future. This data-driven feedback loop is crucial for continuous improvement in engineering design and product lifecycle management.

The application of AI in software engineering and development itself is also profound. AI can assist in code generation, debugging, and testing, making the software development lifecycle more efficient. Tools are emerging that can automatically write boilerplate code, suggest code completions, identify potential bugs, and even generate test cases based on requirements. This frees up software engineers to focus on more complex architectural decisions and innovative problem-solving. For instance, AI-powered code assistants can analyze existing codebases, identify security vulnerabilities, and suggest remediation strategies. Automated testing frameworks that leverage AI can explore a wider range of scenarios and edge cases than manual testing, leading to more robust and secure software. This is particularly important as software becomes increasingly embedded in critical systems, from medical devices to financial trading platforms, where reliability and security are paramount. The ability of AI to analyze patterns in code and identify potential issues before they manifest in production is a significant enhancement to quality assurance and risk mitigation in software engineering.

The integration of AI into engineering design and practice is not merely about adopting new tools; it represents a fundamental evolution in the engineering mindset. It calls for a deeper understanding of data science, machine learning, and computational thinking, alongside traditional engineering disciplines. Engineers are increasingly becoming orchestrators of intelligent systems, leveraging AI to achieve outcomes that were previously unattainable. This shift necessitates a continuous learning approach, as the field of AI is rapidly evolving. The ability to effectively deploy AI in engineering relies on a solid understanding of its strengths and limitations, ensuring that it is applied ethically and responsibly. The goal remains to enhance human problem-solving capabilities, not to supplant them, leading to innovations that are not only technically superior but also more aligned with human needs and values. As we move forward, the symbiotic relationship between human ingenuity and artificial intelligence will define the next generation of engineering breakthroughs.

The burgeoning capabilities of artificial intelligence are undeniably transformative, promising to reshape industries and redefine what is possible. Yet, the trajectory of these powerful technologies is not solely dictated by algorithmic sophistication or computational might. A crucial, and often understated, element in unlocking AI's full potential lies in the deliberate and thoughtful application of human-centered design principles. The conventional wisdom might suggest that AI development is a purely technical endeavor, a domain for computer scientists and data engineers alone. However, this perspective overlooks a fundamental truth: artificial intelligence is ultimately created by humans, for humans, and its impact is felt most profoundly within human contexts. To merely build intelligent systems without deeply considering their users, their ethical implications, and their place in the broader societal fabric is to risk creating powerful tools that are either misaligned with human needs, prone to unintended consequences, or simply fail to achieve their intended purpose.

This is where the synergy between design and AI becomes not just beneficial, but essential. Design thinking, at its core, is an iterative process focused on understanding the user, challenging assumptions, redefining problems, and creating innovative solutions to prototype and test. When applied to AI development, this methodology shifts the focus from a purely technical "can we?" to a more nuanced and impactful "should we?" and "how should we?" It begins with empathy, with a deep dive into the needs, desires, pain points, and workflows of the people who will interact with or be affected by an AI system. This user-centric approach ensures that AI is not developed in a vacuum, but rather as a responsive and supportive element within existing human systems. For example, imagine an AI designed to assist medical professionals in diagnosing diseases. A purely technical approach might focus on maximizing diagnostic accuracy based on vast datasets. While accuracy is paramount, a design-led approach would also consider the physician's workflow, the time constraints of a busy clinic, the need for clear and interpretable explanations from the AI, and the potential for alarm fatigue. The design process would involve observing physicians at work, interviewing them about their challenges, and prototyping AI interfaces that seamlessly integrate into their diagnostic process, providing actionable insights without overwhelming them or eroding their autonomy.

The integration of design thinking into AI development fundamentally reorients the creation process. Instead of viewing AI as an end in itself, it becomes a means to an end – the end being the enhancement of human experience, efficiency, and well-being. This involves moving beyond superficial user interfaces and delving into the deeper aspects of interaction, trust, and the overall user journey. Consider the development of AI-powered educational tools. While the algorithms might be brilliant at tailoring learning paths based on student performance, a design perspective would scrutinize how students actually engage with the material, how they perceive the AI as a learning companion, and how the tool fosters intrinsic motivation and a genuine love for learning. This might lead to incorporating elements of gamification, personalized feedback that goes beyond mere correctness, and interfaces that encourage collaboration rather than isolation. Without this design layer, even the most sophisticated AI might fall short, failing to inspire or effectively engage its intended audience.

Furthermore, the ethical considerations that are increasingly central to discussions about AI are intrinsically linked to design. Ethical AI is not an add-on feature; it must be woven into the very fabric of the development process from its inception. Design plays a critical role in operationalizing ethical principles. For instance, addressing bias in AI is a major challenge. Bias can creep in through biased training data, flawed algorithmic design, or biased human interpretation of

AI outputs. A design-led approach proactively seeks to identify and mitigate these biases. This might involve carefully curating and scrutinizing training datasets for representation, designing algorithms with fairness metrics built-in, and creating user interfaces that allow users to understand potential biases and provide feedback. When developing an AI system for hiring, for example, a design approach would not only ensure the AI's predictive accuracy but also actively work to prevent it from perpetuating historical biases against certain demographic groups. This could involve designing the system to anonymize certain personal characteristics during initial screening, or to flag potential biases in its recommendations for human review.

The concept of explainable AI (XAI) is another area where design and AI synergy is paramount. As AI systems become more complex and their decisions have greater impact, the need for transparency becomes critical. Users, regulators, and developers alike need to understand how an AI arrives at its conclusions. Design thinking can help translate complex algorithmic processes into understandable and actionable explanations. Instead of presenting a black box output, a well-designed AI system will offer context, reasoning, and confidence levels, empowering users to trust and effectively utilize the AI's capabilities. For example, in a financial context, an AI recommending a particular investment strategy would need to not only suggest the strategy but also explain *why* it is being recommended, referencing market trends, risk assessments, and historical performance data in a clear and accessible manner. This makes the AI a partner in decision-making rather than an opaque oracle.

The synergy between design and AI also fosters innovation that is not just technically impressive but also novel and user-valuable. By deeply understanding user needs and contexts, designers can identify unmet needs that AI can then be leveraged to address in creative ways. This often leads to products and services that are not merely incremental improvements but represent genuine breakthroughs. For instance, the development of AI-powered personal assistants, like those found on smartphones and smart speakers, was not solely a feat of natural language processing. It was also a testament to design that anticipated user desires for convenience, information access, and hands-free interaction. The success of these assistants stems from their ability to understand natural language requests, respond helpfully, and integrate with other services, all while maintaining a persona that feels approachable and reliable. This holistic approach, combining AI's intelligence with design's understanding of human interaction, has fundamentally altered how many people engage with technology on a daily basis.

Moreover, the iterative nature of design thinking is particularly well-suited to the evolving landscape of AI. AI models often require continuous refinement and adaptation as new data becomes available and user behaviors change. A design process that incorporates feedback loops and user testing throughout the development lifecycle ensures that AI systems remain relevant, effective, and aligned with user expectations over time. This contrasts with a purely engineering-driven approach, which might focus on achieving a specific performance metric and then consider the product "finished," only to find it quickly becomes outdated or misaligned with user needs in the real world. By embracing design's emphasis on continuous improvement and user feedback, AI development becomes a more dynamic and responsive process, leading to more robust and long-lasting solutions.

The practical implementation of this synergy often involves interdisciplinary teams. AI researchers and engineers work alongside UX/UI designers, ethicists, domain experts, and even social scientists. This collaboration ensures that a diverse range of perspectives is brought to bear on the AI development process. Designers act as translators, bridging the gap between complex technical capabilities and human understanding, ensuring that the AI is not only functional but also usable, desirable, and ethically sound. They champion the user, asking critical questions about accessibility, inclusivity, privacy, and the potential for unintended social consequences. This collaborative model moves away from siloed development and towards a more holistic, integrated approach that acknowledges the multifaceted nature of creating impactful AI.

The multiplicative effect of this synergy cannot be overstated. When AI's power to process information, identify patterns, and automate tasks is combined with design's ability to understand users, anticipate needs, and craft intuitive experiences, the result is systems that are far greater than the sum of their parts. AI provides the engine of intelligence, but design provides the steering wheel, the navigation system, and the comfortable cabin. It ensures that this powerful engine is directed towards beneficial outcomes, that it is accessible to those who need it, and that it operates within ethical and societal boundaries. For instance, AI in healthcare can analyze medical images for early signs of disease, but a design approach ensures that the output is presented to a clinician in a way that is clear, actionable, and integrated into their workflow, thereby improving patient care. Similarly, AI can optimize energy grids for efficiency, but a design approach ensures that the resulting policies and interfaces are fair and do not disproportionately burden certain communities.

In essence, the intersection of design and AI is where technology transcends mere functionality to become truly human-centric. It is about building AI that empowers, that assists, that enhances, and that ultimately serves humanity. This requires a fundamental shift in how we approach AI development – moving from a purely technical pursuit to a discipline that embraces empathy, ethics, and user understanding as core components. The future of impactful AI lies not just in the intelligence of the algorithms, but in the wisdom of their design. By thoughtfully integrating design thinking into every stage of AI creation, we can ensure that these powerful technologies are harnessed responsibly, leading to innovations that are not only intelligent but also humane, and that truly bridge the gap between the potential of machines and the needs of people.

The question of "why now?" is paramount when discussing the urgency of responsible AI. We stand at a unique juncture in technological history, where the theoretical possibilities of artificial intelligence are rapidly materializing into tangible applications that permeate nearly every facet of our lives. This acceleration is not a gradual evolution; it is a seismic shift driven by a confluence of factors that amplify both the potential benefits and the inherent risks of AI. To ignore this urgency is to risk being swept away by a tide of technological advancement without adequate steering or safety measures, potentially leading to outcomes that are detrimental to individuals and society at large.

One of the most compelling reasons for the immediate need for responsible AI is the sheer velocity of its adoption. What were once experimental concepts or niche tools are now being integrated into core infrastructure and everyday consumer products at an unprecedented rate. From sophisticated recommendation engines that curate our online experiences to AI-powered diagnostic tools in healthcare, and from autonomous vehicles navigating our streets to intelligent assistants managing our schedules, AI is no longer a futuristic prospect; it is a present reality. This rapid integration means that the decisions made today about AI development and deployment will have immediate and far-reaching consequences. There is little room for leisurely deliberation when the foundational elements of our digital and physical worlds are being reshaped so quickly. The risk is that we deploy systems that are inadequately understood, poorly governed, or ethically unsound, simply because the pressure to adopt the latest AI capabilities is overwhelming. This rapid pace necessitates a proactive, rather than reactive, approach to responsibility. We cannot afford to wait for widespread harm to occur before we address the ethical implications; by then, the damage may be irreversible, or the embedded systems may be too deeply entrenched to alter easily.

Furthermore, the complexity of contemporary AI systems outpaces traditional oversight mechanisms and even our own intuitive understanding. Modern AI, particularly deep learning models, often operates as a "black box." While we can design them and train them with vast datasets, the intricate web of connections and computations that lead to a specific output can be opaque even to their creators. This inherent complexity makes it challenging to anticipate all potential failure modes or unintended consequences. When an AI system makes a decision – whether it's approving a loan, recommending a course of treatment, or flagging a potential security threat – understanding *why* it made that decision becomes critical. Without this transparency, accountability is elusive, and the ability to correct errors or mitigate bias is severely hampered. The growing intricacy of these systems demands that responsibility be baked into their design from the outset, rather than being an afterthought applied to the opaque output of a complex algorithm. We must move beyond simply celebrating AI's predictive power and focus on its interpretability, fairness, and robustness in diverse real-world scenarios.

This growing complexity and rapid adoption inevitably bring to the forefront a host of potential societal impacts that demand our immediate attention. Concerns about bias are no longer theoretical; they are well-documented and have real-world consequences. AI systems trained on historical data can inadvertently perpetuate and even amplify existing societal biases related to race, gender, socioeconomic status, and other protected characteristics. This can lead to discriminatory outcomes in areas like hiring, criminal justice, and access to services, further entrenching inequality. For example, facial recognition systems have shown differential accuracy rates across demographic groups, and AI used in recruitment has been found to penalize candidates based on keywords associated with gender. The ethical imperative here is clear: we must develop and deploy AI that actively works to counter, rather than reinforce, societal inequities.

Privacy is another critical concern amplified by the pervasive nature of AI. Many AI applications rely on the collection and analysis of vast amounts of personal data. While this data can fuel powerful insights and personalized experiences, it also raises significant questions about data security, consent, and the potential for misuse. As AI systems become more sophisticated in their ability to infer sensitive information from seemingly innocuous data points, the risk of intrusive surveillance and manipulation grows. Ensuring that AI respects individual privacy and adheres to robust data protection principles is not just a legal requirement; it is a fundamental ethical obligation that requires proactive design and policy interventions. The convenience offered by AI should not come at the cost of our fundamental right to privacy.

The economic implications of AI, particularly concerning job displacement and the future of work, also underscore the urgency for responsible development. As AI automates tasks previously performed by humans, there is a legitimate concern about widespread job losses and the widening of economic disparities. While AI can also create new jobs and augment human capabilities, the transition requires careful management to ensure that the benefits are broadly shared and that adequate support systems are in place for those whose livelihoods are affected. Responsible AI development must consider these socioeconomic impacts, advocating for policies and strategies that promote inclusive growth and a just transition to an AI-augmented economy. This involves not only technological innovation but also societal adaptation and proactive planning.

Moreover, there is a growing global awareness and demand for AI to be developed and deployed in alignment with human values. As AI systems become more autonomous and influential, the need to ensure they operate ethically and beneficially is becoming a central theme in public discourse, policy discussions, and academic research. International organizations, governments, and civil society groups are increasingly calling for ethical frameworks, regulatory guidelines, and standards for AI development. This collective awakening signifies that the conversation around AI has moved beyond purely technical feasibility to encompass its profound societal implications. The "now" is urgent because the global community is recognizing the need for guardrails and is actively seeking ways to shape AI's trajectory.

The urgency is also driven by the potential for AI to be used for malicious purposes. While AI offers immense benefits, its power can also be harnessed for surveillance, disinformation campaigns, autonomous weapons, and cyber warfare. The development of powerful AI technologies without robust ethical considerations and safeguards creates vulnerabilities that can be exploited by those with harmful intentions. Therefore, a commitment to responsible AI is not just about maximizing benefits; it is also about mitigating risks and preventing the weaponization of these technologies. This requires a concerted effort from researchers, developers, policymakers, and international bodies to establish norms and controls that prevent AI from being used to undermine peace, security, and human rights.

In essence, the urgency surrounding responsible AI stems from the unprecedented power and reach of these technologies, coupled with their accelerating integration into society. The rapid pace of adoption, the increasing complexity of AI systems, and the growing awareness of their profound societal impacts – including bias, privacy violations, economic disruption, and potential misuse – all converge to create a critical moment. We are at a point where proactive, human-centered design and ethical considerations are not optional enhancements but fundamental requirements. Failing to act now to imbue AI development with responsibility risks creating a future where powerful intelligent systems operate beyond human control, amplify existing societal problems, and undermine the very values we seek to uphold. This chapter, and indeed this book, is a call to action, emphasizing that the time to prioritize responsible AI is not in some distant future, but here and now. The decisions made in the coming years will set the course for how AI shapes our world, and it is imperative that this shaping is guided by wisdom, foresight, and a deep commitment to human well-being.

This book embarks on a comprehensive exploration of the intricate and evolving relationship between humans and machines, with a particular focus on the principles and practices of human-centered artificial intelligence

(AI) within engineering design. As we move from understanding the fundamental shifts in this convergence to actively shaping its future, it becomes essential to chart a clear path through the multifaceted landscape of this discipline. This concluding section of our introductory chapter is designed to serve as precisely that: a roadmap for our collective journey, outlining the structure of the forthcoming discussions and setting the stage for the in-depth analyses and practical insights that lie ahead. Our aim is to equip you, the reader, with a clear understanding of what to expect, how the material builds upon itself, and the ultimate value proposition of engaging with the subsequent chapters.

Our exploration will unfold in a deliberate and progressive manner, beginning with the foundational concepts that underpin the human-machine nexus and gradually advancing towards the sophisticated applications and ethical considerations that define contemporary AI engineering. The subsequent chapters are meticulously structured to guide you from a broad understanding of the present technological milieu to a nuanced appreciation of the challenges and opportunities that lie at the intersection of human ingenuity and artificial intelligence.

The initial segments of this book are dedicated to establishing a robust conceptual framework. We will delve deeply into the historical evolution of human-machine interaction, tracing the arc from early mechanical aids to the sophisticated digital systems of today. This historical perspective is crucial for understanding the trajectory of technological development and the fundamental shifts that have brought us to the current era of intelligent machines. Following this historical context, we will rigorously examine the core principles of artificial intelligence, demystifying key concepts such as machine learning, deep learning, natural language processing, and computer vision. This will not be a purely theoretical exercise; rather, each concept will be elucidated with practical examples and analogies to make them accessible to a broad audience, including those without a deep technical background. The objective here is to ensure a common ground of understanding, equipping you with the essential vocabulary and conceptual tools needed to navigate the more complex discussions that will follow. We will address the fundamental building blocks of AI, exploring how algorithms learn, how data shapes their behavior, and the inherent limitations and biases that can arise from these processes. This foundational knowledge is indispensable for anyone seeking to engage critically and constructively with AI technologies.

As we solidify this understanding, the book will transition into the heart of human-centered AI in engineering design. This pivotal section will explore the methodologies and philosophies that place human needs, values, and experiences at the forefront of AI development. We will investigate the principles of user-centered design as they apply to intelligent systems, focusing on aspects like usability, accessibility, and the creation of intuitive and engaging human-AI interfaces. This involves a deep dive into how engineers and designers can actively solicit and incorporate user feedback throughout the development lifecycle, ensuring that AI systems are not only technically proficient but also genuinely beneficial and adaptable to human users. We will discuss techniques for understanding user behavior, cognitive load, and emotional responses when interacting with AI, and how these insights can be translated into design choices that foster trust and effective collaboration. The concept of "explainable AI" (XAI) will be a significant focus, examining how to make AI decision-making processes transparent and understandable to humans. This is critical for building trust, enabling debugging, and ensuring accountability. We will explore various XAI techniques and their implications for different AI applications, from medical diagnostics to financial services.

Furthermore, the ethical dimensions of AI will be woven throughout these discussions, not as an isolated topic, but as an integral part of responsible engineering. We will address the pervasive issues of bias in AI systems, exploring its sources, its manifestations, and the critical importance of developing strategies for mitigation. This includes examining fairness metrics, debiasing techniques, and the role of diverse datasets and development teams in creating more equitable AI. Privacy concerns will also be a central theme. As AI systems become more adept at collecting and analyzing vast amounts of personal data, understanding and implementing robust privacy-preserving techniques in AI design is paramount. We will discuss concepts like differential privacy, federated learning, and the importance of informed consent and data governance in the AI era. The book will provide practical guidance on how engineers can embed privacy by design into their AI solutions, ensuring that user data is protected and handled ethically.

A significant portion of the book will be dedicated to exploring the practical application of human-centered AI principles across various engineering domains. We will present case studies and examples from fields such as healthcare, transportation, education, manufacturing, and entertainment. These examples will illustrate how human-centered AI can be used to solve real-world problems, enhance human capabilities, and create novel experiences. For instance, in healthcare, we will examine AI-powered diagnostic tools that augment clinicians' abilities, personalized treatment plans, and intelligent systems for patient monitoring, all designed with the needs of both patients and medical professionals in mind. In the realm of autonomous systems, we will discuss the intricate challenges of designing vehicles, robots, and drones that can safely and effectively interact with human environments and operators, emphasizing the importance of robust safety protocols, clear communication interfaces, and fail-safe mechanisms. The manufacturing sector will provide insights into how AI can be used to optimize production processes, improve quality control, and create safer working environments through intelligent automation and human-robot collaboration. In education, we will look at AI tutors and personalized learning platforms that adapt to individual student needs, providing tailored feedback and support while respecting pedagogical principles and the role of human educators.

Beyond specific applications, we will also delve into the broader societal implications of AI. This includes discussions on the future of work, the economic impact of automation, and the importance of reskilling and upskilling the workforce. We will explore how human-centered AI design can contribute to a more equitable distribution of benefits arising from AI adoption, ensuring that technological progress leads to inclusive growth rather than exacerbating inequalities. The role of AI in addressing global challenges, such as climate change, poverty, and disease, will also be highlighted, emphasizing the potential for responsible AI to be a powerful force for good when guided by human values and ethical principles. This involves a careful consideration of how AI can be leveraged to enhance our understanding of complex systems, optimize resource allocation, and facilitate innovative solutions to some of humanity's most pressing problems.

As we progress through the book, the focus will increasingly shift towards the future trajectory of human-machine collaboration. We will explore emerging trends in AI, such as embodied AI, affective computing, and the development of more general forms of artificial intelligence. These discussions will be framed by our core principles of human-centered design and ethical responsibility. The book will encourage a forward-thinking perspective, prompting readers to consider the long-term societal impacts of these advancements and the proactive measures needed to steer AI development in a direction that benefits humanity. This involves fostering a culture of continuous learning and adaptation, as the field of AI is characterized by rapid innovation and evolving challenges. We will discuss the importance of interdisciplinary collaboration, bringing together engineers, ethicists, social scientists, policymakers, and the public to shape the future of AI in a democratic and inclusive manner.

The structure of the book is designed to provide a logical flow of information, building knowledge layer by layer. We begin with the "what" and "why"—understanding the current state of human-machine convergence and the imperative for a human-centered approach. We then move to the "how"—exploring the principles, methodologies, and tools for designing and developing responsible AI systems. Finally, we address the "where" and "what next"—examining practical applications across diverse domains and contemplating the future of human-AI collaboration, always with an eye toward ethical considerations and societal well-being. Each chapter will strive to provide not just theoretical insights but also actionable guidance, offering practical advice, frameworks, and best practices that engineers, designers, and stakeholders can apply in their work.

In essence, this book is intended to be more than a mere academic treatise; it is a call to action, a guide for navigating the complex and exciting terrain of human-centered AI in engineering. By the end of our journey together, you will have a comprehensive understanding of the principles that should guide AI development, the practical techniques for implementing these principles, and the ethical considerations that are paramount for ensuring that AI serves humanity's best interests. The roadmap laid out here is designed to demystify the field, empower you with knowledge, and inspire you to become an active participant in shaping a future where humans and machines collaborate effectively, ethically, and beneficially. We invite you to delve into the subsequent chapters with an open mind and a commitment to building AI systems that augment our capabilities, respect our values, and contribute positively to the world.

Chapter 2: Foundations of Human-Centered AI

Understanding the individuals who will interact with AI systems, and the environments in which these interactions will occur, is the bedrock upon which all successful human-centered AI engineering is built. This foundational step moves beyond merely defining functional requirements; it delves into the rich tapestry of human experience, uncovering the unspoken needs, implicit assumptions, and situational nuances that can profoundly influence the adoption, efficacy, and overall impact of an AI solution. To truly center AI on humans, we must first commit to a rigorous and empathetic exploration of who these humans are, what they are trying to achieve, and the multifaceted contexts in which they operate. This exploration is not a perfunctory checkbox; it is an ongoing, iterative process that informs every subsequent design and development decision.

The methodologies employed in this phase of user understanding draw heavily from the broader field of human-computer interaction (HCI) and design research. Among the most powerful techniques is **ethnography**, a qualitative research method that involves immersing the researcher in the user's natural environment to observe their behaviors, interactions, and challenges firsthand. Unlike a laboratory setting, ethnography captures the authentic, often unarticulated, aspects of user behavior that arise from the interplay of technology, tasks, social dynamics, and the physical surroundings. For instance, imagine designing an AI-powered scheduling assistant for busy medical professionals. An ethnographic study might involve shadowing doctors and nurses through their demanding shifts, observing not just how they currently manage their schedules but also the interruptions they face, the informal communication channels they rely on, the impact of emergency situations on their plans, and the emotional toll of constant time pressure. This kind of deep, observational understanding can reveal critical needs that a simple survey might miss – perhaps the need for an AI that can dynamically re-optimize schedules based on real-time patient status changes, or an interface that can be operated with minimal visual attention during critical procedures. Without this contextual immersion, an engineer might mistakenly design a system that, while technically sophisticated, is impractical or even disruptive in the high-stakes environment of a hospital.

Complementing ethnographic observation are **in-depth interviews**. These are structured or semi-structured conversations designed to elicit detailed information about users' experiences, motivations, pain points, and aspirations. When conducted effectively, interviews go beyond surface-level questions to probe deeper into the 'why' behind user actions and preferences. For an AI designed to assist individuals with managing chronic health conditions, an interview might explore not just their current medication adherence practices but also their feelings of empowerment or disempowerment, their understanding of their condition, their social support networks, and their anxieties about long-term health. An AI developer seeking to understand the needs of elderly users for an AI-driven home safety system would move beyond asking "Do you want a safety alert system?" to exploring their daily routines, their concerns about falling, their desire for independence, their comfort level with technology, and their existing social connections. Questions like "What are the biggest challenges you face in maintaining your independence at home?" or "Can you describe a time you felt particularly vulnerable in your home?" can unlock crucial insights that inform the design of an AI that is not only functional but also supportive and respectful of the user's autonomy and dignity. The interviewer's skill lies in creating a safe space for participants to share openly, actively listening, and employing probing techniques to uncover latent needs.

Surveys and questionnaires, while often perceived as more quantitative, can also play a significant role in understanding user needs, particularly when used to gather data from a larger user base or to validate findings from qualitative research. However, for human-centered AI, the design of these instruments is paramount. Open-ended questions are essential for capturing qualitative richness, and carefully crafted multiple-choice options can help to quantify specific preferences or behaviors identified through other methods. For example, a survey for an AI-powered financial planning tool might include questions about users' current financial literacy levels, their attitudes towards risk, their primary financial goals (e.g., saving for a down payment, retirement, debt reduction), and their preferred communication styles for receiving financial advice. The phrasing of these questions must be clear, unbiased, and framed from the user's perspective. A question like "How often do you review your investment portfolio?" is less informative than "How confident do you feel about making informed decisions regarding your investments, and what steps do you currently take to manage them?"

A crucial element that underpins all these research methods is
empathy. This is the ability to understand and share the feelings of another, to step into their shoes and perceive the world from their vantage point. In the context of AI design, empathy is not merely a desirable trait; it is a necessary skill. Engineers and designers must cultivate a deep sense of empathy to truly grasp the user's experience, particularly when dealing with individuals or situations that differ significantly from their own. This might involve designing AI for individuals with disabilities, where understanding the physical, cognitive, and social barriers they face is critical. For example, an AI-powered communication aid for someone with a speech impairment requires more than just speech recognition; it necessitates an understanding of their frustration with misinterpretations, their desire to express complex thoughts, and the importance of non-verbal cues in communication. Empathy guides the designer to ask not just "How can this AI understand their speech?" but "How can this AI empower them to communicate effectively and authentically, preserving their dignity and self-expression?"

Empathy also helps in recognizing and addressing the

diverse range of user capabilities and backgrounds. Not all users will have the same technical proficiency, cognitive abilities, or cultural perspectives. An AI system designed for a global audience, for instance, must consider linguistic diversity, varying levels of digital literacy, and cultural norms around privacy, authority, and decision-making. An AI tutor for young children will require a vastly different approach - in terms of interface, language, and interaction style - than an AI knowledge management system for experienced researchers. The empathetic designer anticipates these differences and actively seeks to understand the unique challenges and preferences of each user group.

The concept of
latent needs is particularly important in user research for AI. These are needs that users may not be consciously aware of, or are unable to articulate, but which become evident through observation and careful analysis of their behavior. Often, users adapt their workflows or develop workarounds to compensate for system limitations without realizing that a better solution is possible. For example, users of complex data analysis software might spend hours manually cleaning and formatting data before it can be processed by an AI model. They may not explicitly state "I need better data pre-processing tools," but their significant time investment in this manual task clearly indicates a latent need. An AI engineering team that identifies this latent need can then develop AI-powered data wrangling features that automate these tedious tasks, freeing up users to focus on higher-value analytical work.

Similarly, in the context of AI for creative professionals, users might be accustomed to laborious iteration cycles for generating design variations. An empathetic observation might reveal the underlying desire for more intuitive and rapid exploration of creative possibilities, a latent need that can be addressed by AI-driven generative design tools.

Contextual factors are the environmental, social, and situational elements that surround user interaction with an AI system. These factors can significantly influence how an AI is used, perceived, and its ultimate effectiveness. Understanding these contexts is as vital as understanding the user themselves. Consider the example of an AI-powered diagnostic tool for radiologists. The context here is not just the radiology reading room; it's the pressure of high caseloads, the potential for fatigue, the need for rapid decision-making, the reliance on precise visual interpretation, and the ethical imperative of accurate diagnosis. An AI that is designed without considering these contextual pressures might be visually cluttered, require too many steps to operate, or fail to provide clear, actionable insights in a time-sensitive manner. Conversely, an AI that is designed with an awareness of these factors might offer streamlined workflows, highlight critical findings with appropriate visual cues, and provide confidence scores or explanations for its recommendations, all of which are tailored to the demands of the radiologist's work environment.

The context might also be social. An AI designed for a

collaborative team needs to understand the group dynamics, communication patterns, and decision-making processes within that team. An AI for an elder care setting involves not only the elderly individual but also their caregivers, family members, and medical professionals. Understanding the relationships, communication flows, and potential conflicts or synergies between these different stakeholders is crucial for designing an AI that integrates smoothly into the care ecosystem. For an AI that assists in emergency response, the context is one of extreme stress, limited communication bandwidth, and life-or-death decisions. The AI must be robust, reliable, and provide information in a format that is easily digestible and actionable under immense pressure.

Furthermore, the physical environment plays a significant role. An AI integrated into a manufacturing plant must contend with noise, dust, vibration, and potentially hazardous conditions. An AI used in a retail environment might need to adapt to variable lighting, crowded spaces, and the presence of both customers and staff. The design of the AI's interface, its form factor (if it has a physical embodiment), and its interaction modalities must all be appropriate for these physical realities. A voice-based AI in a noisy factory floor might be impractical, whereas a wearable display or a system that uses haptic feedback could be far more effective.

The process of understanding user needs and context is not a one-time event but a continuous loop. Initial research provides a foundation, but as the AI system is designed, prototyped, and tested, new insights emerge. Feedback from early prototypes can reveal unforeseen usability issues, misinterpretations of user needs, or novel contextual challenges. This iterative feedback loop, driven by ongoing user engagement and contextual awareness, is what distinguishes human-centered AI from traditional engineering approaches. It ensures that the AI system evolves in alignment with the human experience it is intended to serve, leading to solutions that are not only technically sound but also deeply relevant, usable, and beneficial. It is through this diligent, empathetic, and context-aware approach that we can truly harness the power of AI to augment human capabilities and improve lives.

The advent of artificial intelligence presents a profound opportunity to redefine what it means for a technology to be "human-centered." In the preceding discussion, we established that at its core, human-centered design prioritizes understanding the user and their context. However, as AI systems become more sophisticated, capable of learning, adapting, and even exhibiting emergent behaviors, our conception of human-centeredness must evolve to address a more nuanced set of challenges and ethical imperatives. It is no longer sufficient to merely ensure an AI is usable or efficient; we must also consider its impact on human autonomy, cognitive load, emotional states, and ultimately, its broader societal footprint.

Moving beyond the fundamental principles of human-computer interaction, human-centered AI necessitates a deep consideration of how these intelligent systems either augment or, conversely, diminish human capabilities and agency. The goal is not to simply build AI that performs tasks *for* humans, but rather AI that collaborates *with* humans in a way that enhances their own skills, judgment, and decision-making power. This means designing AI systems that empower users, rather than supplanting their critical thinking or decision-making authority. For instance, an AI-powered diagnostic tool in healthcare, while incredibly powerful in identifying patterns invisible to the human eye, should be designed to assist a physician's judgment, not replace it. It should present its findings with a degree of uncertainty, offer supporting evidence, and explain its reasoning, thereby enabling the physician to make a more informed decision. If the AI were to present a diagnosis as an absolute certainty, it would erode the physician's autonomy and critical role, potentially leading to over-reliance and a degradation of their diagnostic skills over time. This principle extends to virtually every domain where AI is deployed.

A truly human-centered AI system is one that inherently respects and upholds user control. This means users should, wherever feasible, have the ability to understand, influence, and override the AI's actions or recommendations. This is particularly critical in AI systems that make decisions with significant consequences, such as in loan applications, hiring processes, or even personalized content delivery. For example, a financial AI that automatically reallocates a user's investments based on perceived market trends should provide clear mechanisms for the user to review these changes, understand the rationale behind them, and manually adjust or reverse them if they disagree. The absence of such control can lead to a feeling of powerlessness and a lack of trust in the system, undermining its long-term adoption and user satisfaction. This concept of user control is intertwined with the fundamental human need for agency – the capacity to act independently and make choices that shape one's own life. AI systems that systematically remove this agency, even with the best intentions of efficiency, risk diminishing the human experience.

Transparency, another cornerstone of human-centered AI, is essential for fostering trust and enabling meaningful user control. In the context of AI, transparency refers to the degree to which the inner workings and decision-making processes of the system are understandable to the user. This is a complex challenge, especially with advanced machine learning models that can operate as "black boxes." However, even in the absence of complete explainability for highly complex models, efforts must be made to provide meaningful transparency. This could involve revealing the types of data used for training, the general principles guiding the AI's operation, or providing simplified explanations for specific outputs. For example, a content recommendation AI might not reveal the exact algorithmic formula used to select a video, but it could transparently state that it recommends content based on viewing history, likes, and the popularity of similar videos among users with similar tastes. This level of transparency allows users to understand *why* certain content is being presented and provides them with context to make informed choices about their engagement. When AI operates opaquely, users are left guessing, fostering suspicion and a sense of being manipulated.

Furthermore, human-centered AI must actively guard against manipulative or coercive designs. This is a particularly thorny ethical area, as AI's ability to understand and predict human behavior can be leveraged to subtly influence decisions in ways that may not be in the user's best interest. Consider the "dark patterns" that have emerged in user interface design, where elements are intentionally crafted to trick users into taking unintended actions. AI can amplify these manipulative tactics. For instance, an AI-powered e-commerce system could, by analyzing a user's browsing behavior and emotional state (inferred from their interactions), subtly nudge them towards impulse purchases by presenting limited-time offers or creating a false sense of scarcity. A human-centered approach would dictate that the AI's persuasive capabilities are used ethically, respecting the user's autonomy and aiming to provide genuine value, rather than exploiting psychological vulnerabilities. This requires designers to develop a robust ethical compass and implement safeguards to prevent the misuse of AI for undue influence. The goal should always be to empower users to make informed choices, not to coerce them into actions they might later regret.

The impact of AI on human cognitive load is another crucial aspect of human-centered design. While AI can automate complex tasks and reduce mental effort, it can also introduce new forms of cognitive burden. For example, an AI system that generates an overwhelming number of alerts or notifications can lead to "alert fatigue," where users become desensitized and may miss critical information. Conversely, an AI that requires extensive configuration or troubleshooting can place a heavy cognitive load on users who are not technically expert. Human-centered AI strives for an optimal balance, offloading cognitive effort where beneficial without overwhelming the user or creating new mental demands. This might involve intelligent filtering of information, providing clear and concise summaries, or designing intuitive interfaces that minimize the need for complex decision-making during operation. The aim is to make the interaction with the AI feel effortless and supportive, allowing users to focus their cognitive resources on higher-level tasks.

Emotional well-being is also a significant, and often overlooked, dimension of human-centered AI. AI systems can evoke a range of emotional responses, from delight and satisfaction to frustration, anxiety, and even anger. An AI that is unreliable, buggy, or difficult to use can lead to significant emotional distress. Conversely, an AI that is empathetic, responsive, and helpful can foster positive emotions and a sense of well-being. This is particularly important in AI applications designed for sensitive domains like mental health support, elder care, or education. For instance, an AI companion for individuals experiencing loneliness needs to be programmed with a nuanced understanding of human emotional needs, offering comfort and support without overpromising or creating unrealistic expectations. The design should prioritize creating a sense of connection and understanding, while also being mindful of the potential for emotional dependence or disappointment. This requires designers to consider the affective dimensions of human-AI interaction and to engineer systems that are not only functional but also emotionally intelligent and supportive.

The societal impact of AI systems is a broader, yet equally critical, concern for human-centered design. While individual usability and user experience are paramount, we must also consider the collective implications of widespread AI adoption. This includes issues of fairness, bias, equity, and the potential for AI to exacerbate existing societal inequalities. A human-centered approach demands that AI developers actively work to identify and mitigate biases in their systems, ensuring that they do not discriminate against particular demographic groups. For example, an AI used for criminal justice risk assessment must be scrutinized for racial or socioeconomic biases that could lead to unfair sentencing. Similarly, AI systems deployed in hiring processes must be carefully designed to avoid perpetuating gender or racial discrimination.

Moreover, human-centered AI must consider the

broader economic and social transformations that AI will bring. This includes the impact on employment, the digital divide, and the potential for increased surveillance or control. While not all these issues can be solved solely through design, a human-centered ethos compels us to ask critical questions about the long-term consequences of our creations. Are we designing AI that will create new jobs and opportunities, or will it lead to widespread unemployment? Will AI bridge the digital divide, or will it further marginalize those who lack access to technology? By proactively addressing these societal questions, human-centered AI development can contribute to a more equitable and beneficial future for all.

Ultimately, the objective of defining "human-centered" in the AI era is to establish a robust ethical framework for AI design and development. This framework goes beyond mere technical specifications and functional requirements, extending into the realm of human values, rights, and well-being. It is a commitment to building AI that serves humanity, rather than the other way around. This means prioritizing user autonomy, ensuring meaningful control, championing transparency, and actively guarding against manipulation and coercion. It involves a deep consideration of AI's impact on cognitive load and emotional states, and a proactive engagement with the broader societal implications of these powerful technologies.

This comprehensive view of human-centeredness recognizes that AI is not just a tool; it is a force that can shape human experiences, relationships, and societies. Therefore, its design and deployment must be guided by a profound respect for human dignity, agency, and flourishing. The development of AI should not be a race to create the most powerful or intelligent systems, but rather a conscientious effort to create systems that are wise, ethical, and ultimately, beneficial to humanity. This requires a multidisciplinary approach, bringing together engineers, designers, ethicists, social scientists, and users themselves to co-create a future where AI truly augments human potential and contributes to a more just and equitable world. The ethical considerations we have outlined – autonomy, transparency, control, and the mitigation of negative impacts – are not add-ons; they are integral to the very definition of what it means for AI to be genuinely human-centered. This ethical grounding provides the necessary compass for navigating the complex landscape of AI development and ensuring that its immense power is wielded for the good of all.

In navigating the intricate landscape of Artificial Intelligence design, a robust ethical framework is not merely an option but a fundamental necessity. As AI systems permeate virtually every facet of modern life, their potential to both uplift and undermine human well-being necessitates a set of guiding principles that ensure their development and deployment are aligned with human values. These principles serve as our moral compass, directing us toward creating AI that is beneficial, just, and respectful of human dignity.

At the forefront of this ethical imperative is the principle of
fairness. In the context of AI, fairness primarily addresses the insidious issue of algorithmic bias. AI systems learn from data, and if the data they are trained on reflects existing societal biases, the AI will inevitably perpetuate and even amplify these inequities. This can manifest in discriminatory outcomes across various domains, from loan applications and hiring decisions to criminal justice risk assessments and healthcare diagnostics. For instance, an AI used for hiring that has been trained on historical data where men disproportionately held certain positions might unfairly disadvantage female applicants, not because of their qualifications, but because the algorithm has learned to associate success with gender, mirroring past societal prejudices. Ensuring fairness requires meticulous attention to data collection, preprocessing, and model evaluation. It involves actively identifying and mitigating bias, employing techniques such as data augmentation, re-weighting samples, or using bias-aware algorithms. The consequences of neglecting fairness are profound: the erosion of trust, the perpetuation of systemic discrimination, and the creation of a technologically-enabled underclass. A fair AI system is one that treats all individuals and groups equitably, irrespective of their background or characteristics, and strives to rectify, rather than reinforce, historical injustices. This principle demands ongoing vigilance, as biases can be subtle and evolve

over time, necessitating continuous monitoring and adjustment of AI systems.

Closely linked to fairness is the principle of **accountability**. When an AI system makes an error, causes harm, or produces an undesirable outcome, it is crucial to establish who is responsible. In traditional engineering, accountability is often straightforward – the engineer, the manufacturer, or the operator is held liable. However, the autonomous and self-learning nature of AI complicates this. Is it the data scientists who trained the model, the developers who coded the system, the company that deployed it, or perhaps the AI itself (a notion that raises complex legal and philosophical questions)? Establishing clear lines of accountability is paramount for building trust and ensuring that recourse is available when things go wrong. This involves creating mechanisms for oversight, audit trails, and clear decision-making hierarchies. For example, if an autonomous vehicle causes an accident, investigations must be able to trace the decision-making process of the AI, the parameters it operated under, and the responsibilities of the various human actors involved in its creation and deployment. Without accountability, there is little incentive to rigorously test, validate, and monitor AI systems, leading to a higher risk of catastrophic failures and a reluctance for individuals and organizations to adopt AI technologies. Accountability fosters responsibility and incentivizes the development of robust, safe, and reliable AI systems, ultimately protecting individuals and society from harm.

Transparency is another bedrock principle, which, as discussed previously, refers to the understandability of an AI system's operations and decision-making processes. While achieving full explainability for highly complex deep learning models remains a significant challenge, a commitment to transparency means striving for clarity to the greatest extent possible. This involves making the data used for training, the algorithms employed, and the rationale behind specific outputs understandable to relevant stakeholders. The level of transparency required may vary depending on the application; for a simple recommendation engine, a general explanation of how recommendations are generated might suffice. However, for AI systems making high-stakes decisions, such as those in healthcare or finance, a more detailed explanation of the reasoning process, including confidence levels and potential sources of error, is essential. For instance, an AI diagnostic tool that suggests a particular course of treatment should be able to explain *why* it arrived at that recommendation, highlighting the symptoms, test results, and medical literature that support its conclusion. This allows medical professionals to critically evaluate the AI's suggestion and make an informed judgment, rather than blindly accepting an opaque output. Lack of transparency breeds suspicion and erodes trust. Users and stakeholders need to understand, at some level, how an AI system functions to be confident in its reliability and fairness. It also enables better debugging, auditing, and the identification of unintended consequences.

The principle of
privacy is intrinsically linked to the data that fuels AI systems. AI, particularly machine learning, often relies on vast amounts of data, much of which can be personal and sensitive. Protecting individual privacy involves ensuring that data is collected, used, and stored responsibly and ethically. This means obtaining informed consent for data collection, anonymizing or de-identifying data wherever possible, and implementing robust security measures to prevent data breaches. AI systems should be designed with privacy-preserving techniques such as differential privacy, federated learning, or homomorphic encryption, which allow models to be trained or used without directly accessing or exposing sensitive raw data. For example, a smart home device that uses AI to learn user habits should do so in a way that localizes processing as much as possible, or anonymizes the data before sending it to the cloud, and clearly informs the user about what data is being collected and how it is used. The consequences of neglecting privacy are severe: identity theft, reputational damage, and a chilling effect on freedom of expression and association as individuals fear pervasive surveillance. Upholding privacy rights is not just a legal obligation but a moral imperative that respects individual autonomy and dignity in the digital age.

Security is a crucial, yet often conflated, aspect of AI ethics. While privacy focuses on protecting personal information from unauthorized access or misuse, security pertains to safeguarding the AI system itself from malicious attacks and ensuring its integrity and reliability. AI systems can be vulnerable to various forms of attack, such as adversarial attacks, where subtle perturbations to input data can cause the AI to misclassify or make incorrect decisions. For example, a slight alteration to an image that is imperceptible to the human eye could cause an AI image recognition system to identify a stop sign as a speed limit sign, with potentially fatal consequences in an autonomous driving context. Robust security measures are needed to defend against such attacks, including techniques for detecting and mitigating adversarial examples, securing training data, and protecting the AI model from unauthorized modification or access. Furthermore, ensuring the security of AI systems means guaranteeing their availability and resilience in the face of potential disruptions. Neglecting security can lead to system failures, data breaches, manipulation of AI outputs for malicious purposes, and a loss of public trust in AI technologies. A secure AI system is one that is resilient to attack, operates reliably, and protects its data and functionality from compromise.

Finally, the principle of **beneficence** compels us to design AI systems that actively contribute to human well-being and societal good. This means moving beyond simply avoiding harm and actively seeking to create AI that solves pressing global challenges, improves quality of life, and empowers individuals. This could involve developing AI for medical research to discover new cures, creating AI tools to combat climate change by optimizing energy consumption, or designing AI that provides personalized education and support to learners of all ages and abilities. For instance, an AI developed to assist in disaster response could analyze satellite imagery to identify areas in need of immediate aid, optimize logistical routes for rescue teams, and predict potential secondary hazards. The pursuit of beneficence requires foresight and a deep understanding of human needs and societal aspirations. It means intentionally directing AI development towards applications that yield positive outcomes for humanity, rather than solely focusing on commercial interests or technical novelty. The ethical imperative here is to ensure that the immense power of AI is harnessed for the betterment of humankind, fostering progress, equity, and prosperity. It requires a conscious effort to identify opportunities where AI can make a meaningful difference and to invest in research and development that aligns with these goals.

These principles – fairness, accountability, transparency, privacy, security, and beneficence – are not isolated ideals but are deeply interconnected. A system that is secure is more likely to protect privacy. Transparency is essential for accountability and for ensuring fairness. Beneficence guides the application of all other principles, ensuring that AI is developed and used for good. Embracing these ethical principles is not an impediment to innovation; rather, it is the foundation for sustainable and responsible AI development. It is through this ethical lens that we can ensure that AI truly serves humanity, augmenting our capabilities, enhancing our lives, and contributing to a future that is not only technologically advanced but also just, equitable, and humane. This ethical scaffolding is crucial for guiding engineers, designers, policymakers, and users alike as we continue to integrate these powerful systems into the fabric of our society. The responsible creation of AI hinges on a steadfast commitment to these values, ensuring that as AI evolves, it does so in a manner that is aligned with our deepest human aspirations.

The integration of Artificial Intelligence into our lives is often framed through a lens of automation, promising to take over tasks previously performed by humans. However, a more nuanced and, arguably, more impactful perspective views AI not as a replacement, but as a powerful amplifier of human capabilities. This approach champions AI as a sophisticated partner, designed to augment our innate skills, extend our cognitive reach, and elevate our decision-making processes. Instead of a zero-sum game where AI gains and human roles diminish, this perspective envisions a synergistic collaboration, where human and artificial intelligence work in concert to achieve outcomes far exceeding what either could accomplish alone.

Consider the realm of complex data analysis. Humans are adept at pattern recognition, intuitive leaps, and understanding context. However, when confronted with datasets of astronomical proportions – encompassing petabytes of information generated by scientific experiments, financial markets, or sensor networks – even the most brilliant human mind can struggle to discern meaningful signals from the noise. This is where AI truly shines as an augmentation tool. Machine learning algorithms can process these vast datasets with unparalleled speed and accuracy, identifying subtle correlations, anomalies, and trends that might remain hidden to human observation. For instance, in climate science, AI can analyze meteorological data, satellite imagery, and historical records to predict weather patterns with greater precision, identify the early signs of extreme weather events, or model the complex interactions within ecosystems. This doesn't replace the climatologist; rather, it equips them with a more powerful lens through which to interpret the world, allowing them to focus their expertise on strategic planning, policy recommendations, and communicating complex findings. The AI handles the heavy lifting of data processing, freeing up the human expert to engage in higher-order thinking and strategic action.

Similarly, in the demanding field of surgery, AI is emerging as an invaluable co-pilot. Robotic-assisted surgery has already revolutionized precision and minimally invasive techniques. Augmenting this with AI takes it a step further. Imagine an AI system integrated into the surgical suite, analyzing real-time imaging (such as MRI or CT scans) alongside the surgeon's movements and vital signs of the patient. This AI can provide crucial, real-time information to the surgeon, highlighting critical anatomical structures to avoid, predicting potential complications based on subtle physiological changes, or even suggesting optimal instrument trajectories. It acts as an intelligent assistant, constantly monitoring, analyzing, and providing relevant data precisely when and where it's needed. The surgeon's years of training, their tactile sense, and their critical judgment remain paramount, but the AI enhances their situational awareness and provides an additional layer of precision and safety. This collaboration can lead to shorter operation times, reduced patient trauma, and improved recovery rates – outcomes that are a direct result of human expertise being amplified by intelligent technological support.

The domain of research is another fertile ground for AI-driven augmentation. For academic researchers, sifting through the ever-expanding volume of scientific literature, experimental data, and theoretical models is a monumental task. AI-powered tools can drastically accelerate this process. Natural Language Processing (NLP) algorithms can scan, categorize, and summarize thousands of research papers, identifying key findings, methodologies, and potential research gaps. AI can also analyze experimental data, such as genomic sequences, protein structures, or particle collision results, to identify promising avenues for further investigation or to uncover unexpected relationships. For example, in drug discovery, AI can screen millions of chemical compounds, predicting their potential efficacy and side effects, thereby significantly narrowing down the candidates for laboratory testing. This dramatically speeds up the discovery pipeline, allowing researchers to focus their efforts on the most promising leads and to delve deeper into the scientific questions at hand. The AI acts as an tireless research assistant, uncovering patterns and insights that a human researcher might take years to find, if they found them at all.

This augmentation extends to everyday professional tasks as well. Customer service representatives, for instance, can benefit from AI assistants that provide them with instant access to relevant customer history, product information, or troubleshooting guides. As a customer describes their issue, the AI can proactively surface the most pertinent information, allowing the representative to provide faster, more accurate, and more personalized support. This doesn't diminish the human interaction; in fact, it can enhance it by freeing the representative from the burden of searching for information, allowing them to focus on active listening, empathy, and problem-solving. The AI handles the retrieval and organization of data, enabling the human to excel at the relational and empathetic aspects of their role.

Even in creative fields, AI can serve as a powerful augmentative tool. While the notion of AI replacing human creativity is often debated, a more constructive application involves AI as a muse or a collaborator. AI-powered design tools can generate a multitude of visual options based on user-defined parameters, providing designers with a diverse palette of ideas to draw from. For writers, AI can assist with grammar and style checking, suggest alternative phrasing, or even generate plot outlines based on genre conventions. In music composition, AI can generate harmonic progressions or melodic ideas that a composer can then refine and build upon. The AI here is not the artist, but a sophisticated tool that can overcome creative blocks, offer novel perspectives, and expand the range of possibilities, allowing the human artist to focus on their unique vision and artistic intent.

The key to successful AI augmentation lies in the design philosophy. Instead of aiming for full automation, the goal is to create AI systems that seamlessly integrate with human workflows, respecting human judgment and expertise. This requires careful consideration of the user interface, the types of information the AI provides, and the degree of control the human operator retains. The AI should be designed to present information in an easily digestible format, offering explanations and justifications for its suggestions or analyses, thereby fostering trust and allowing for critical evaluation. For example, an AI system designed to assist in financial trading might flag a particular stock as a potential investment opportunity. It should also provide the underlying reasons for this flagging – perhaps based on market trends, company performance indicators, or news sentiment analysis. The human trader can then review this information, apply their own market knowledge and risk appetite, and make the final decision. The AI has augmented their decision-making by highlighting a possibility they might have overlooked, but the ultimate responsibility and strategic insight remain with the human.

Furthermore, the concept of "explainable AI" (XAI) is particularly relevant in the context of augmentation. When an AI system is augmenting human decision-making, especially in high-stakes environments like healthcare or aviation, it is crucial that humans understand
why the AI is making a particular recommendation or flagging a certain piece of information. Transparency in AI decision-making builds confidence and allows for the identification and correction of errors. An AI diagnostic tool that suggests a particular treatment, for instance, should be able to articulate the symptoms, test results, and medical literature that informed its suggestion. This allows the physician to understand the AI's reasoning, compare it with their own clinical judgment, and make a more informed and confident decision. The AI has augmented the physician's diagnostic process by offering a data-driven perspective, but the final medical judgment rests with the human expert, supported by a clear understanding of the AI's input.

The development of AI as an augmentation tool also necessitates a shift in training and education. As AI becomes more integrated into various professions, individuals will need to learn how to effectively collaborate with these systems. This includes understanding the capabilities and limitations of AI, developing skills in interpreting AI-generated insights, and mastering the interfaces through which humans and AI interact. Educational institutions and professional organizations will play a vital role in equipping the workforce with the necessary competencies to thrive in an AI-augmented future. This isn't about replacing human skills with AI, but about cultivating new hybrid skills that leverage the strengths of both human and artificial intelligence.

The ethical implications of AI augmentation are also significant. While the focus is on enhancing human capabilities, it's crucial to ensure that these augmentation tools do not inadvertently lead to deskilling, over-reliance, or the erosion of human judgment. The design of AI systems must prioritize maintaining human agency and control. The goal is to empower individuals, not to create dependencies that diminish their inherent abilities. This requires ongoing dialogue and research into the long-term societal impacts of widespread AI augmentation, ensuring that these technologies serve to uplift humanity rather than subtly undermine it. For instance, if an AI system consistently makes better financial decisions than a human advisor, there's a risk that human advisors might become less skilled in their own right over time. This is where the principle of beneficence becomes crucial – ensuring that augmentation leads to genuine improvements in well-being and not unintended negative consequences.

In conclusion, the role of AI in augmenting human capabilities represents a powerful and positive trajectory for technological development. By designing AI systems that act as intelligent assistants, data processors, and insightful collaborators, we can unlock unprecedented levels of human performance, creativity, and understanding. This synergistic approach, where AI amplifies human strengths rather than replacing human roles, offers a pathway towards solving complex global challenges and enriching individual lives. The focus remains squarely on human-centered AI, where technology serves as a tool to extend our reach, deepen our knowledge, and ultimately, to make us more capable, more informed, and more effective in navigating the complexities of the modern world. This requires a conscious and deliberate design effort, a commitment to ethical development, and a forward-looking approach to education and training, ensuring that AI augmentation leads to a future where human potential is not just preserved, but profoundly expanded.

In embarking on any initiative, particularly one as potentially transformative as an Artificial Intelligence (AI) project, the initial act of defining purpose and boundaries is not merely a procedural step; it is the bedrock upon which success is built. This is especially true when adopting a human-centered approach, where the ultimate aim is to develop AI systems that genuinely serve human needs and aspirations, rather than simply pursuing technological advancement for its own sake. The clarity of our goals and the meticulous definition of our scope are the guiding stars that will navigate us through the complexities of AI development, ensuring that our efforts remain anchored in human value.

The fundamental question that must precede any technical specification or algorithmic design is: what specific human problem are we seeking to solve? This is not a question about what AI *can* do, but rather what it *should* do from the perspective of the people it is intended to assist. A human-centered AI project begins with empathy and a deep understanding of the challenges, pain points, and unmet needs experienced by individuals or groups. For instance, instead of aiming to "build a predictive model for customer churn," a human-centered framing might be "help customer service representatives proactively identify and support customers who are at risk of leaving, ensuring they feel valued and receive timely assistance." This reframing shifts the focus from a purely technical output to a tangible improvement in the customer experience and the empowerment of the human agent.

Identifying the intended user groups is equally critical. Who are the individuals that will interact with, benefit from, or be affected by this AI? Understanding their existing workflows, their levels of technical literacy, their cultural contexts, and their unique requirements is paramount. Are we designing for experienced professionals in a highly specialized domain, or for the general public with diverse backgrounds? For example, an AI designed to assist medical professionals in diagnosing rare diseases will require a different interface, a different level of explanatory detail, and different validation mechanisms than an AI designed to help consumers manage their personal finances. Failing to deeply understand the users can lead to the development of AI systems that are either too complex to be adopted, too simplistic to be useful, or even alienating. This user-centric identification informs every subsequent design decision, from the user interface and user experience to the underlying data and the performance metrics.

Beyond identifying the 'what' and the 'who,' a rigorous definition of the 'where' and 'how' – the boundaries of the AI's application – is essential. Scope creep, the uncontrolled expansion of project requirements, is a notorious pitfall in any development endeavor, but it poses unique challenges for AI projects due to their inherent complexity and rapid evolution. Clearly delineating what the AI *will* do and, just as importantly, what it *will not* do, sets realistic expectations and provides a framework for decision-making. This involves defining the specific tasks the AI is designed to perform, the types of data it will process, the environments in which it will operate, and the level of autonomy it will possess. For instance, if an AI is being developed to assist in content moderation on a social media platform, its scope might be defined to flag potentially violating content based on pre-defined rules and machine learning models, but explicitly exclude making final judgment calls or handling nuanced ethical dilemmas that require human discretion.

Setting realistic expectations regarding the capabilities and limitations of the AI is a cornerstone of a successful human-centered project. It is crucial to avoid overpromising and underdelivering, which can erode trust and lead to disappointment. This means acknowledging that AI, even at its most advanced, is not infallible. It can make mistakes, it can exhibit biases inherited from its training data, and its performance can degrade over time if not properly maintained. For a human-centered AI, transparency about these limitations is not a weakness, but a strength. For example, an AI-powered legal research assistant might be highly effective at identifying relevant case law, but it should be clearly communicated that it cannot provide legal advice or substitute for the judgment of a qualified legal professional. This honesty fosters a more mature and effective human-AI collaboration, where users understand how to best leverage the AI's strengths while compensating for its weaknesses.

The process of defining goals and scope is not a one-time event; it is an iterative and collaborative undertaking. It requires active engagement with all stakeholders – the development team, the intended users, domain experts, and even ethicists and policymakers. This dialogue helps to uncover potential blind spots, anticipate unforeseen consequences, and ensure that the project remains aligned with its human-centric objectives. For example, during the definition phase of an AI system designed to optimize traffic flow in a city, initial goals might focus on reducing commute times. However, through stakeholder consultation, it might become apparent that the system must also consider factors like air quality, pedestrian safety, and equitable access to transportation for different communities. These considerations, identified early, can be incorporated into the scope and goals, leading to a more holistic and beneficial AI solution.

Furthermore, the scope definition must consider the entire lifecycle of the AI system. This includes not only the initial development and deployment but also ongoing maintenance, updates, and eventual decommissioning. How will the AI's performance be monitored? How will it be retrained or updated to adapt to changing conditions or new data? What are the plans for addressing emergent biases or unintended consequences? A human-centered approach demands that these long-term considerations are addressed proactively, rather than being treated as afterthoughts. For instance, an AI system designed to assist in hiring decisions must have a clear plan for periodic audits to ensure it is not perpetuating or exacerbating existing societal inequalities, and the scope should include the resources and processes for conducting such audits.

The commitment to a well-defined scope also acts as a crucial safeguard against the temptation to pursue technically elegant but humanly irrelevant features. In the fast-paced world of AI research and development, there is an ever-present allure to chase cutting-edge algorithms or to add functionalities that, while impressive, do not directly contribute to solving the identified human problem. A clear scope provides a filter, helping the team to critically assess new ideas and prioritize those that genuinely enhance the human value proposition. If a proposed feature does not directly support the core goals or serve the intended users effectively, it can be consciously deferred or rejected, maintaining the project's focus and efficiency.

This disciplined approach to goal and scope definition is not about stifling innovation; rather, it is about channeling innovation in a direction that is meaningful and beneficial. It ensures that the significant resources – time, talent, and capital – invested in AI projects are directed towards creating technology that genuinely improves human lives, empowers individuals, and addresses real-world challenges. It establishes a clear north star, guiding the team through the inevitable complexities and uncertainties of AI development, and laying a robust foundation for creating AI systems that are not just intelligent, but also wise, ethical, and truly human-centered. Without this foundational clarity, even the most technically sophisticated AI project risks becoming a solution in search of a problem, or worse, a misdirected effort with unintended negative consequences. The diligent, empathetic, and collaborative work of defining goals and scope is, therefore, the indispensable first step in the journey of human-centered AI.

Chapter 3: Participatory Design in AI Development

The preceding discussion has underscored the critical importance of clearly defined goals and scope in human-centered AI development. It establishes the foundational premise that AI should not be an end in itself, but a tool meticulously crafted to address specific human needs and challenges. This foundational clarity, however, is not an abstract exercise conducted in isolation by a development team. Instead, it is the fertile ground upon which the true power of user involvement blossoms. Moving beyond mere theoretical consideration, the active, consistent, and meaningful inclusion of end-users throughout the AI development lifecycle transforms the very nature of how we conceive, build, and deploy intelligent systems. This subsection delves into the profound advantages of this participatory approach, illustrating how it shifts the paradigm from users as passive recipients of technology to active co-creators of solutions that genuinely resonate with their lived experiences.

The most immediate and perhaps most significant benefit of user involvement is the unparalleled depth of understanding it cultivates regarding user needs, contexts, and aspirations. Traditional methods of gathering user feedback, such as surveys, interviews, or even usability testing at the end of a development cycle, often provide a retrospective or reactive view. They tell us what users think about something that already exists or has been largely conceived. Participatory design, conversely, is inherently proactive. By bringing users into the development process from its nascent stages, we unlock direct access to their tacit knowledge, their unspoken frustrations, and their often-unarticulated desires. Imagine developing an AI-powered tool for agricultural farmers to optimize irrigation. While engineers might focus on sensor data, soil moisture levels, and weather forecasts, a participating farmer can immediately highlight practical considerations such as the variability of field conditions due to microclimates, the physical limitations of operating machinery in certain terrains, or the social dynamics of sharing water resources with neighboring farms. These are nuances that are exceedingly difficult to glean from a survey or a focus group; they emerge organically through direct dialogue and collaborative problem-solving with the people who live and breathe the problem every day. This direct immersion into the user's world significantly reduces the risk of building AI systems that are technically sophisticated but practically irrelevant or even unusable. It acts as an early warning system against the pitfalls of assumptions made by designers and developers who may not share the same lived realities.

This deep understanding, fostered by genuine user participation, directly translates into a significant reduction in the risk of developing AI solutions that miss the mark. The history of technology is replete with examples of innovations that failed not because of technical shortcomings, but because they were not designed with a true appreciation for how people would actually use them, or if they were needed at all. For AI, the stakes are even higher, given the potential for profound societal impact and the significant investment of resources. When users are actively involved in defining the problem, shaping the requirements, and providing feedback on prototypes, the likelihood of creating a system that aligns with their workflows, their cognitive processes, and their ultimate objectives increases exponentially. Consider an AI designed to assist paralegals in legal research. Without user involvement, the system might be programmed to retrieve vast quantities of case law, inundating the paralegal with data without effective categorization or prioritization. A participating paralegal, however, can guide the AI's development by explaining how they typically approach research queries, what types of precedents are most valuable, and what level of detail is required for different stages of a case. This co-creation ensures that the AI becomes a powerful augment to their existing skills, rather than a cumbersome addition to their workload. It moves the AI from being a speculative technological marvel to an indispensable, tailored tool.

Furthermore, the act of involving users as active participants, rather than as mere sources of feedback, fundamentally fosters user buy-in and cultivates trust. When individuals see their needs, ideas, and concerns directly influencing the design and functionality of an AI system, they develop a sense of ownership. This is particularly crucial for AI systems that may bring about significant changes to established practices or decision-making processes. For example, if an AI is being developed to assist managers in resource allocation within a company, involving the managers in the design process – allowing them to shape how the AI presents data, what parameters it considers, and how it flags potential issues – will make them far more likely to adopt and trust the system. They will have a clearer understanding of its underlying logic (to the extent that is possible and appropriate for transparency) and will feel that the system has been built to support, rather than dictate, their professional judgment. This contrasts sharply with a situation where a system is imposed upon them without their input, which can breed suspicion, resistance, and ultimately, failure to achieve the intended benefits. Trust in AI is not an inherent quality of the technology itself; it is built through transparency, reliability, and a clear demonstration that the system is designed with the user's best interests at heart. Participatory design is a potent mechanism for building this trust, as it inherently demonstrates a commitment to user-centricity.

The paradigm shift from traditional user feedback to active participation is a critical aspect of this subsection. Historically, user feedback often involved providing opinions on a finished or near-finished product. This might manifest as bug reports, feature requests submitted after deployment, or usability testing sessions where users are asked to perform predefined tasks on a working system. While valuable, this approach is limited because it occurs late in the development cycle, making significant design changes costly and time-consuming. Moreover, it often elicits reactions to existing design choices rather than generative input that can shape the core direction of the

AI. Participatory design, on the other hand, elevates users to the role of co-creators. They are not just testers; they are collaborators who contribute to the conceptualization, the specification, the prototyping, and the iterative refinement of the AI. This can take many forms: user workshops where requirements are brainstormed and prioritized, co-design sessions where users sketch out interface ideas, or even scenarios where users actively contribute to the labeling or curation of data that will train the AI. For instance, in developing an AI tutor for students learning a new language, participatory methods might involve students working alongside developers to define the types of conversational practice they find most helpful, the error correction strategies that are most encouraging, and the progression of learning modules that best suits their pace. This collaborative co-creation ensures that the AI is not merely a reflection of the developers' understanding of learning, but a dynamic system built in partnership with the learners themselves.

This collaborative approach is indispensable for ensuring that AI solutions are truly tailored to the people they are intended to serve, making the technology more effective and accepted. When users are involved as co-creators, the AI system becomes an extension of their needs and capabilities, rather than an external imposition. This tailoring leads to AI that is not only more useful but also more equitable. For example, an AI designed to assist individuals with disabilities in navigating public transportation needs to be developed with direct input from those individuals. Their unique

challenges – from understanding audio announcements to operating touch-screen interfaces with limited dexterity – must inform every aspect of the AI's design, from its voice output and button sizes to its information hierarchy and error handling. Without this direct participation, the AI could inadvertently create new barriers or fail to address existing ones. By contrast, an AI developed through a participatory process is inherently more likely to be inclusive and effective. It moves beyond a one-size-fits-all approach to technology and embraces the diversity of human experience. This ensures that the AI does not just function, but that it functions
for the people it is meant to help, in ways that are meaningful and impactful to their lives, fostering adoption and maximizing the positive potential of artificial intelligence.

In the preceding discussion, we established the imperative of user involvement in AI development, shifting from a model of passive reception to active co-creation. This foundational shift is not merely about gathering feedback; it's about embedding user intelligence, context, and lived experience into the very fabric of AI design. Having explored the advantages of this participatory ethos – the enhanced understanding, risk mitigation, and fostered trust – we now transition to the practical methodologies that bring this principle to life. Specifically, this section will delve into the concrete techniques for participatory data collection, a crucial phase where users actively contribute to the raw material that fuels AI models. This is where the abstract notion of co-creation becomes tangible, through methods that empower users and ensure the ethical, privacy-preserving, and representative nature of the data gathered.

The process of data collection for AI development has historically been a somewhat opaque affair, often conducted by technical teams without direct or meaningful engagement with the end-users. This can lead to datasets that are biased, incomplete, or do not accurately reflect the real-world variability and nuances of the problems AI is intended to solve. Participatory data collection directly addresses this by inviting users to become active participants in generating and curating the data that will train and evaluate AI systems. This approach not only enriches the dataset but also imbues the entire process with a greater degree of transparency and user agency. It moves beyond simply asking users to *use* an AI, to asking them to help *build* the intelligence that underpins it.

One of the most direct and effective methods for participatory data collection is through **user-led data labeling workshops**. In these structured sessions, participants are guided through the process of annotating data relevant to the AI's intended function. For example, if the AI is designed to identify different types of wildlife from camera trap images, users who are familiar with local fauna can be invited to label images with the correct species. This is far more nuanced than having a generic labeling service perform the task, as local experts can identify subtle distinctions, regional variations in appearance, or even behaviors that might be indicative of specific species. The workshop setting allows for immediate clarification of guidelines, iterative refinement of annotation criteria, and importantly, a collaborative environment where users can discuss edge cases and arrive at a consensus. This not only improves the quality and accuracy of the labeled data but also provides developers with invaluable qualitative insights into the decision-making processes of the users, which can inform further AI development and interface design. The key here is not just to collect labels, but to understand the reasoning behind them. For instance, in a workshop for an AI designed to detect crop diseases, farmers might not only label images of diseased plants but also explain the environmental conditions under which the disease typically appears, the stages of plant growth most susceptible, or the treatments they have historically found effective. This rich, contextual information, gleaned directly from the participants, is often more valuable than the labels themselves and can guide the AI's feature engineering and predictive capabilities.

Beyond labeling,
co-design sessions focused on feature definition
offer another powerful avenue for participatory data collection. In this context, users collaborate with developers to identify what aspects of the data are most relevant and meaningful for the AI's task. For an AI intended to assist urban planners in predicting traffic congestion, for instance, commuters could be involved in defining what constitutes a "significant congestion event," what types of data points are most indicative of impending jams (e.g., specific road segments, times of day, weather events, public transport disruptions), and

what thresholds of speed or density are critical. These sessions often involve brainstorming, sketching, and prioritizing potential data features, ensuring that the AI is trained on metrics that genuinely reflect the lived experience of traffic and not just abstract sensor readings. This collaborative feature engineering is crucial because users, immersed in the problem domain, often possess an intuitive understanding of causal relationships and important indicators that developers might overlook. They can articulate the "why" behind certain traffic patterns – the impact of school drop-off times, the ripple effect of a minor accident, or the influence of local events – which can then be translated into data features. This goes beyond merely identifying data sources; it involves shaping the very conceptualization of what data matters.

A particularly impactful approach is **in-situ data gathering**, where users contribute data within their natural environments and workflows. This method is invaluable for capturing the dynamic and context-dependent nature of many real-world phenomena. For example, patients participating in the development of a personalized health AI could be encouraged to log symptoms, medication adherence, or daily activities directly through a user-friendly mobile app as they experience them. This avoids the recall bias inherent in retrospective surveys and captures data in its most authentic form. Similarly, for an AI designed to optimize energy consumption in smart homes, residents could provide data by interacting with smart devices and sharing their energy usage patterns and preferences. The crucial element of in-situ gathering is its seamless integration into daily life, minimizing the burden on participants while maximizing the ecological validity of the data. Developers must ensure that the tools used for in-situ collection are unobtrusive, intuitive, and provide clear value to the user, perhaps through personalized insights or feedback generated by the AI based on the data they contribute. This symbiotic relationship, where user contribution directly leads to user benefit, is a powerful motivator for sustained participation. For instance, an app designed for citizen scientists to report air quality could prompt users to record measurements along with contextual information like their proximity to busy roads, the presence of industrial activity, or their subjective experience of air quality. This data, collected "in the wild," provides a richer picture than fixed monitoring stations alone, and by sharing aggregated, anonymized findings back with the community, the app demonstrates its utility and fosters continued engagement.

Crucially, all participatory data collection methods must be underpinned by robust **ethical and privacy-preserving frameworks**. Empowering users to contribute data necessitates a commitment to transparency, consent, and control. Users must be fully informed about what data is being collected, how it will be used, who will have access to it, and what measures are in place to protect their privacy and security. This often involves employing techniques such as data anonymization, pseudonymization, differential privacy, and federated learning, where appropriate. Consent should be granular, allowing users to opt in or out of specific data collection activities, and they should retain the right to access, modify, or delete their data. For instance, when collecting sensitive health data, stringent protocols must be in place to ensure compliance with regulations like HIPAA or GDPR. Patients should have clear visibility into which AI models are being trained on their data and the ability to revoke their consent at any time, with the assurance that their data will be removed from future training sets. Furthermore, developers should actively consider how to mitigate inherent biases in user-contributed data. While users bring invaluable lived experience, their contributions can still reflect societal inequalities or specific demographic profiles. Techniques like stratified sampling, bias detection algorithms, and ongoing audits of data representativeness are essential to ensure that the AI developed is fair and equitable.

Consider the development of an AI navigation app for cyclists. Participatory data collection could involve cyclists using a specialized app to log their routes, perceived safety of road segments, road surface conditions, and even incidents encountered. To ensure privacy, routes could be anonymized, and personal identifiers removed. Cyclists might be given the option to report incidents anonymously, with robust data validation to prevent misuse. The app could then provide personalized route suggestions based on aggregated data of other cyclists' experiences, highlighting safer or more pleasant paths. The co-design aspect comes in when cyclists help define what constitutes a "safe" or "pleasant" route – is it the presence of bike lanes, the absence of heavy traffic, the smoothness of the road surface, or the availability of scenic views? This iterative feedback loop, where users contribute data and then see tangible improvements informed by that data, builds trust and encourages ongoing participation. The ethical considerations here would include ensuring that data isn't used to profile individuals or predict their movements in ways that could compromise their safety or privacy.

Another compelling example is the development of an AI system to assist in disaster response coordination. Participatory data collection could involve emergency responders, community leaders, and affected individuals contributing real-time information about needs, resource availability, and damage assessments. This might be facilitated through a secure, offline-capable mobile application. For instance, community members could report critical needs like access to clean water or medical supplies, while responders could log the location and status of damaged infrastructure or the deployment of aid. The ethical imperative here is paramount: ensuring the data is handled with the utmost confidentiality, is accessible only to authorized personnel, and is used solely for the purpose of improving response efforts. Privacy-preserving techniques are essential, especially when dealing with vulnerable populations. Anonymizing reports of individual hardship while still capturing the *type* and *location* of need, for example, allows for effective resource allocation without compromising individual privacy. Participatory design in this context means that the interface and the types of information collected are designed *with* the people who will be using it in high-stress, potentially chaotic environments. This ensures the data is actionable, reliable, and collected efficiently when time is of the essence. The system should also provide feedback to the community about how their contributions are being used, fostering a sense of collective effort and shared responsibility in disaster recovery.

The integration of **crowdsourcing platforms** for data collection can also be a powerful participatory tool, provided it is managed with care and ethical considerations at the forefront. These platforms can scale data collection efforts significantly by engaging a large number of users in tasks that might be too small or too numerous for dedicated workshop sessions. For an AI designed to monitor biodiversity through acoustic recordings, for instance, a crowdsourcing platform could allow users to upload short audio clips and label them with potential animal sounds. Advanced users could be invited to contribute more detailed annotations, such as the specific type of bird call or the presence of other ambient noises. The ethical considerations for crowdsourcing involve fair compensation for contributors, clear data usage policies, and mechanisms to ensure data quality and prevent malicious input. Transparency regarding how the collected data will be used to train AI models is critical. For example, if users are contributing to an AI that identifies invasive species, they should be informed that their labels are helping to build a system that aids in conservation efforts. The platform should also include quality control mechanisms, such as having multiple users label the same data point and using consensus to determine the final label, or employing AI models to flag potentially incorrect annotations for human review. This layered approach ensures that the vast amounts of data collected through crowdsourcing are both useful and reliable, while respecting the contributions of the individuals involved.

Furthermore, the concept of **"data donation"** represents a more advanced stage of user empowerment. Here, individuals or organizations voluntarily contribute datasets they have already collected, with the explicit purpose of advancing AI research and development in a particular domain. This could range from academic researchers sharing anonymized experimental data to individuals contributing their personal data, such as fitness tracker logs or photographic archives, for specific AI projects. The key to successful data donation lies in clear data governance, robust anonymization, and an explicit understanding of the intended use cases. For example, a hospital might decide to donate anonymized patient records to an AI research institution working on early disease detection. This requires stringent ethical review, de-identification protocols that go beyond simple pseudonymization, and a clear agreement on how the data will be secured and used. Patients whose data is included should ideally have been informed and consented to their data being used in such a manner, even if anonymized. Similarly, an individual might choose to donate their collection of nature photography to train an AI for species identification, with the understanding that the AI will be used for conservation and education. The ethical framework around data donation must prioritize user autonomy, ensuring that consent is informed and revocable, and that the data is used in ways that align with the donor's intentions and societal good, preventing its misuse for commercial exploitation or discriminatory purposes.

In essence, the transition to participatory data collection marks a profound shift in how we conceive of and build AI. It recognizes that the most valuable data is often embedded within human experience, context, and expertise. By developing and employing methods like user-led labeling workshops, co-designed feature definition, in-situ data gathering, ethical crowdsourcing, and structured data donation, we can move beyond data collection as a purely technical task to one that is deeply collaborative, ethical, and user-centric. This not only leads to more accurate and effective AI models but also fosters greater trust, transparency, and equity in the development and deployment of artificial intelligence. The examples provided, from agricultural farmers and urban commuters to patients and emergency responders, illustrate the diverse applicability and transformative potential of these participatory approaches. By empowering users to become active participants in the data lifecycle, we ensure that AI is developed not just
for people, but *with* them, leading to technologies that are more aligned with human needs and values.
The preceding sections have underscored the fundamental shift from viewing users as passive recipients of AI to active co-creators, emphasizing their

vital role in shaping the data that fuels these intelligent systems. This foundational principle of involving users in the genesis of AI extends beyond data collection to the very core of how users will interact with and experience these technologies. Now, we pivot to a crucial phase: **co-designing AI interactions**. This involves a deliberate and collaborative process where users, developers, and designers work hand-in-hand to sculpt the conversational dynamics, behavioral patterns, and visual interfaces of AI systems. The objective is to move beyond simply understanding user needs to actively building AI experiences that resonate with human cognition, preferences, and expectations, thereby fostering a sense of intuitive understanding, control, and ultimately, trust.

When we speak of co-designing AI interactions, we are not referring to traditional user interface (UI) or user experience (UX) design alone, although these disciplines are integral. Instead, we are addressing the unique challenges and opportunities presented by AI's emergent capabilities, its potential for proactivity, and its ability to learn and adapt. AI interactions are often more dynamic and less predictable than those with conventional software. An AI might initiate a conversation, offer unsolicited advice, or perform actions based on inferred intent rather than explicit commands. Co-designing these nuanced interactions requires a deep dive into how users conceptualize intelligence, how they communicate with others, and what they anticipate from a helpful assistant. It's about translating the abstract intelligence of an AI into a tangible, comprehensible, and trustworthy encounter.

One of the most effective methodologies for co-designing AI interactions is **collaborative user journey mapping**. This goes beyond standard journey mapping, which typically visualizes a user's experience with a product or service. In a co-design context, the journey map becomes a shared canvas where users and designers jointly chart not only the user's path but also the AI's potential points of engagement, its anticipated actions, and the emotional or cognitive states of the user at each juncture. For instance, when developing an AI-powered mental wellness companion app, a co-design workshop might begin with users describing their personal journeys through periods of stress or anxiety. As they articulate their experiences, needs, and coping mechanisms, the design team, alongside the users, collaboratively sketches out how the AI could interject, offer support, or guide them through exercises. This might involve defining moments where the AI could proactively check in, suggest a mindfulness exercise based on detected speech patterns indicating stress, or offer a listening ear. The "mapping" is not just about where the user goes, but also about how the AI anticipates and responds within that journey. Users can directly contribute to defining the AI's tone of voice, the timing of its interventions, and the types of supportive language it employs. This shared visualization helps to uncover potential misalignments between user expectations and AI behavior early in the design process, facilitating a more user-aligned AI personality and interaction style. The map becomes a living document, annotated and refined collaboratively, ensuring that the AI's presence feels supportive rather than intrusive or uncanny.

Expanding on this, **collaborative interface sketching and prototyping** are critical for making abstract AI interactions concrete. This is where users actively contribute to the visual and functional design of the AI's interface. Instead of developers presenting a finished UI for feedback, users are invited to sketch their ideal interfaces, indicate where they would expect certain information to appear, and describe how they would want to interact with AI-generated suggestions or responses. For an AI that assists with complex data analysis, users might not only point out where they'd like to see visualizations but also sketch out interactive elements that allow them to drill down into the data or adjust parameters. These sketches, often rough and hand-drawn, are incredibly rich in conveying user mental models. They reveal assumptions about how information should be organized, how controls should function, and what level of detail is necessary. Designers and developers can then translate these user-generated ideas into more polished prototypes, which are then brought back to the users for further iteration. This iterative sketching and prototyping cycle ensures that the AI's interface is not just aesthetically pleasing but deeply functional and aligned with how users naturally think about and operate within a given domain. For example, when designing an AI assistant for a busy kitchen, chefs might sketch out how they would want to receive recipe instructions – perhaps as spoken prompts with visual cues on a screen that they can easily glance at while their hands are occupied. They might also sketch out how they would verbally acknowledge steps or ask for clarification, directly influencing the design of voice command interfaces and feedback mechanisms. The act of sketching together fosters a sense of shared ownership over the final product and helps to prevent the "developer knows best" mentality that can lead to unintuitive AI systems.

The conversational nature of many AI systems necessitates a particular focus on **co-designing conversational flows and dialogue**. This involves users and designers working together to script the back-and-forth exchanges between a human and the AI. This is more than just writing dialogue; it's about understanding the pragmatics of human conversation – turn-taking, indirectness, politeness, and the management of ambiguity. For an AI customer service agent, for instance, users might play out common customer service scenarios, articulating how they would express frustration, ask clarifying questions, or confirm understanding. Designers can then work with these participants to craft AI responses that are empathetic, clear, and efficient. This could involve developing a range of pre-defined responses that can be dynamically assembled based on the context, or training the AI to understand and generate more flexible dialogue. Users can also help identify potential points of misunderstanding or "failure modes" in the conversation, where the AI might misinterpret the user or provide an unhelpful response. Collaborative dialogue mapping sessions can be employed, where participants map out different conversational paths, including graceful ways for the AI to recover from errors or to ask for clarification when it's unsure. For example, if an AI is designed to help users troubleshoot a technical issue, users could provide examples of how they would phrase problems, what jargon they would use, and what kinds of troubleshooting steps they would expect to be guided through. They could also articulate how they would feel if the AI didn't understand them and what kind of reassurances would be helpful. This iterative process of dialogue scripting, testing, and refining with users is essential for building AI conversational agents that feel natural and helpful, rather than robotic or frustrating.

A particularly insightful technique for co-designing AI interactions is
role-playing AI behaviors. This method encourages participants to embody both the user and the AI, stepping into each other's shoes to understand the dynamics of the interaction from both perspectives. In a role-playing session for an AI that assists in creative writing, for example, one group of participants might act as the writer, describing their creative block or their ideas, while another group acts as the AI, offering prompts, suggesting plot twists, or providing stylistic feedback. By actively performing these roles, users can directly experience how the AI's suggestions feel – are they inspiring, intrusive, or confusing? This hands-on approach allows for immediate feedback on the AI's personality, its level of assertiveness, and the helpfulness of its contributions. Developers and designers can observe these role-playing sessions to

gain deep insights into the nuances of the desired AI persona and interaction style. This method is invaluable for exploring the more speculative aspects of AI interaction, such as how an AI might offer encouragement, provide constructive criticism, or even playfully challenge a user's assumptions. For instance, when designing an AI tutor for medical students, role-playing could explore how the AI should gently correct an incorrect diagnosis, provide additional resources when a student struggles with a concept, or offer positive reinforcement for correct reasoning. The act of embodying the AI allows users to articulate what kind of "personality" would be most conducive to learning, making the AI feel like a supportive mentor rather than an impersonal grading machine.

Beyond specific interactions, co-designing AI extends to defining the
AI's perceived agency and proactivity. How much initiative should the AI take? When should it interrupt? How should it signal its intentions? These are critical design decisions that profoundly impact user experience. Through co-design workshops, users can explore and define these boundaries. For an AI system designed to manage a smart home, users might express preferences for how the AI should adjust lighting or temperature – should it learn their habits and adjust automatically, or should it always ask for confirmation? Should it proactively suggest energy-saving measures, or wait to be asked? Collaborative scenario planning can help uncover these preferences. Participants can be presented with various hypothetical situations (e.g., "You're having guests over," "There's a storm

approaching," "You're trying to sleep") and asked to describe how they would want the AI to respond. This collaborative exploration helps to establish an appropriate level of AI autonomy that aligns with user comfort and control. For instance, in an AI-powered personal finance manager, users might decide that while the AI should proactively flag unusual spending patterns, they would prefer to be asked before the AI makes any automated adjustments to their savings or investment allocations. This balance between proactive assistance and user control is a delicate one, best navigated through direct user input. The design of these proactive capabilities is a key area where co-design can prevent AI from feeling overbearing or, conversely, too passive to be useful.

Furthermore,
co-designing AI explainability and transparency is crucial for building user trust. Users need to understand, at least at a high level, *why* an AI is behaving in a certain way or making a particular recommendation. This doesn't necessarily mean delving into complex algorithms, but rather providing digestible explanations that align with user understanding. Through co-design, users can indicate what kinds of explanations they find most helpful and at what point in the interaction they would want to receive them. For an AI that recommends personalized news articles, users might express a desire to know *why* a particular article was recommended – was it based on their reading history, topics they've shown interest in, or articles popular among similar readers? Co-design sessions can involve users sketching out where they would expect to

see these explanations appear, what level of detail they would prefer, and how they would want to request more information if needed. This collaborative approach ensures that explainability features are not an afterthought but are integrated seamlessly into the user experience, making the AI feel more transparent and less like a "black box." For an AI diagnostic tool in healthcare, for example, patients and clinicians might collaborate to define how the AI should present its diagnostic possibilities, along with the evidence or key indicators that led to those conclusions. This empowers users to engage critically with the AI's output and fosters confidence in its recommendations.

The application of these co-design principles is evident across a wide spectrum of AI development. Consider an AI designed to assist teachers in grading essays. Instead of developers solely determining how the AI flags grammatical errors or assesses creativity, teachers would be brought into the design process. Through collaborative sketching, they might indicate where they want to see AI-generated feedback appear on a student's paper – perhaps as annotations directly in the margin, or as a summary report. Role-playing sessions could explore how the AI should phrase its suggestions for improvement; teachers would articulate what kind of tone is most effective for motivating students versus pointing out mistakes. They could also co-design the interface for overriding or modifying the AI's grading, ensuring that the human teacher retains ultimate control and can inject their nuanced pedagogical judgment. The AI's "conversational flow" for providing feedback would be shaped by teachers' experience with student learning, ensuring it's constructive and actionable.

Another example can be found in the development of AI assistants for scientific research. Researchers using an AI that helps to analyze experimental data might collaborate with designers to map out their typical research workflows. They would collectively identify points where the AI could offer assistance, such as suggesting relevant literature, identifying potential confounding variables, or predicting the outcome of an experiment based on early results. Collaborative interface sketching would determine how these suggestions are presented – perhaps as pop-up notifications, side panels, or integrated into data visualization tools. Role-playing could explore how the AI should interact when it encounters unexpected or anomalous data; researchers would articulate whether they prefer a cautious "are you sure?" prompt, or a more direct "this data deviates significantly from expectations." The AI's "agency" in suggesting research directions would be carefully negotiated, ensuring it augments rather than dictates the research process. Explainability would be paramount, with researchers co-designing the AI's ability to cite the specific data points or established scientific principles that underpin its suggestions.

In the realm of accessibility, co-designing AI interactions is particularly transformative. For individuals with visual impairments, an AI assistant that reads out screen content needs to be designed with their unique needs in mind. Through collaborative journey mapping, visually impaired users can guide designers through their typical digital tasks, highlighting moments where an AI's audio descriptions are most critical. Interface sketching would involve users dictating where and how they want the AI to provide information – for instance, whether it should announce the type of interactive element (button, link, form field) before reading its label, or provide spatial context for on-screen elements. Role-playing sessions could explore how the AI should handle errors or ambiguities in auditory feedback, with users articulating their preferences for clarification strategies. The AI's "proactivity" might be carefully calibrated; for example, an AI that offers to read an incoming email might do so only after a user explicitly enables that function, respecting their desire for control over auditory interruptions. Co-design ensures that the AI's interaction patterns are not just functional but also empowering and respect the user's sensory experience and autonomy.

Ultimately, the co-design of AI interactions is an iterative and deeply human-centered process. It acknowledges that AI systems are not merely tools but are becoming increasingly like collaborators, companions, and extensions of ourselves. By actively involving users in shaping the AI's interface, its conversational style, its decision-making logic, and its very personality, we move towards building AI that is not only intelligent but also empathetic,

understandable, and aligned with our values. This collaborative approach fosters a sense of ownership, reduces the cognitive load associated with learning new technologies, and builds the essential foundation of trust that is paramount for the widespread and ethical adoption of artificial intelligence. The journey from data co-creation to interaction co-design represents a holistic commitment to developing AI
with people, not just *for* them.
The integration of participatory design principles into AI development brings forth a spectrum of ethical considerations that demand careful attention and proactive management. As we empower individuals and communities to become active contributors in the creation of AI systems, it is imperative to establish robust frameworks that safeguard their rights, respect their contributions, and prevent potential harms. This subsection delves into these critical ethical dimensions, moving beyond the technicalities of co-creation to address the moral underpinnings of building AI *with* people, rather than solely *for* them.

At the forefront of these ethical considerations is the principle of
informed consent, particularly concerning the data users contribute. Participatory AI models often rely on data generated by users, whether through direct input, usage patterns, or feedback loops. For consent to be truly informed, individuals must possess a clear and comprehensive understanding of how their data will be collected, processed, stored, and utilized by the AI system and its developers. This extends beyond a simple acknowledgment checkbox; it requires

accessible and understandable explanations of the AI's purpose, its potential applications, and the specific ways in which their contributions will be used to train, refine, or operate the AI. For example, in the development of a community-based AI that helps manage local resources, consent forms should explicitly detail how data regarding local weather patterns, resource consumption, or community event participation will be anonymized and aggregated, and what types of AI-driven decisions might be informed by this data. Vague or overly technical language can obscure the true implications of data sharing, rendering consent superficial and potentially exploitative. Developers must invest in clear communication strategies, utilizing plain language, visual aids, and even interactive dialogues to ensure participants grasp the full scope of their data's journey.

A related ethical imperative is ensuring **fair compensation and recognition** for the valuable contributions made by participants. When individuals dedicate their time, expertise, and data to the development of an AI system, their efforts should be appropriately acknowledged and, where applicable, compensated. This is particularly pertinent in projects that aim for commercial viability or significant societal impact. The nature of this compensation can vary widely. It might involve direct financial remuneration, access to premium features of the AI system, public acknowledgment through credits or authorship, or even a stake in the intellectual property or profits generated by the AI, especially in community-led initiatives. Consider an AI that assists in medical diagnosis. If clinicians contribute their expertise and anonymized patient data to train this AI, their involvement should be recognized. This could manifest as co-authorship on any resulting publications, inclusion in advisory boards for future development, or even a share of licensing revenue if the AI is commercialized. Conversely, failing to acknowledge or compensate participants can lead to feelings of exploitation and erode trust, undermining the very collaborative spirit that participatory AI seeks to foster. The challenge lies in establishing fair and equitable models that are tailored to the specific context of the AI project and the nature of the contributions.

The question of
privacy and data security is paramount in any AI development, but it takes on heightened significance in participatory contexts. When users are actively engaged in providing data, developers have a heightened responsibility to protect that data from unauthorized access, breaches, or misuse. This involves implementing robust security measures, adopting data minimization practices (collecting only what is strictly necessary), and ensuring that data is anonymized or de-identified where possible and appropriate. For instance, an AI designed to support personalized learning for students requires sensitive data about their academic performance and learning styles. Developers must ensure that this data is stored securely, accessible only to authorized personnel, and that participants (students, parents, or educators) are fully informed about the privacy safeguards in place. Furthermore, they must guard against secondary uses of the data that

were not part of the original consent agreement. The potential for data aggregation and re-identification, even from seemingly anonymized datasets, necessitates ongoing vigilance and a commitment to best practices in data protection.

Moreover, participatory AI development necessitates mechanisms for
recourse and accountability when issues arise. Users who contribute to AI systems must have clear pathways to address concerns, report problems, or seek redress if they believe their rights have been violated or if the AI system causes harm. This could involve establishing dedicated feedback channels, user advocacy groups, or independent oversight committees. For an AI system developed for a specific demographic, such as an AI that assists elderly individuals with daily tasks, it is crucial to have a straightforward process for users or their caregivers to report any instances of the AI behaving inappropriately, causing distress, or malfunctioning in a way that leads to negative consequences. The system should also outline who is accountable for addressing these issues and what remedies are available. Without clear recourse mechanisms, users may feel disempowered and hesitant to participate, and potential harms may go unaddressed, further eroding trust in AI development processes.

A critical ethical consideration, particularly relevant to community-based AI projects, is the
prevention of undue influence and exploitation. When engaging with communities, especially those that may be marginalized or vulnerable, developers must be

mindful of power dynamics and ensure that participation is voluntary and uncoerced. This means avoiding situations where individuals feel pressured to contribute data due to perceived benefits or a lack of alternative options. For example, an AI project aimed at improving public services within an underserved community must ensure that participation is framed as an opportunity, not an obligation, and that the community members' concerns and priorities are genuinely central to the design process. Developers should actively work to identify and mitigate any potential burdens associated with data sharing, such as time commitments or the potential for data to be used in ways that disadvantage the community. This requires genuine partnership, where community members have a voice not only in contributing data but also in shaping the objectives and governance of the AI project itself.

Beyond data contribution, ethical considerations extend to the
representation and inclusivity of participants in the participatory process itself. AI systems are often designed to serve diverse populations, and their development processes must reflect this diversity. This means actively seeking out and including voices from all relevant demographic groups, ensuring that no segment of the intended user base is disproportionately excluded or underrepresented. For an AI designed to assist in legal proceedings, for instance, it is crucial to involve legal professionals from various backgrounds, individuals with different levels of technological literacy, and representatives of diverse socioeconomic and cultural groups. Failure to do so can lead to AI

systems that perpetuate existing biases, overlook critical needs, or even actively disadvantage certain groups. Participatory design efforts must be mindful of accessibility, ensuring that all potential participants can engage meaningfully, regardless of their abilities or technological proficiency. This might involve providing multiple modes of participation, offering training and support, and adapting communication styles to suit different needs.

Furthermore, the
long-term implications and potential for misuse of participatory AI must be considered. As AI systems evolve and become more integrated into society, the data and designs contributed by individuals can have far-reaching consequences. Developers have an ethical responsibility to anticipate and, where possible, mitigate potential negative externalities. This includes thinking about how the AI could be used for surveillance, manipulation, or other harmful purposes, and building in safeguards or ethical guidelines to prevent such outcomes. For an AI that analyzes social media sentiment for marketing purposes, developers must consider the ethical implications of potentially enabling manipulative advertising or the spread of misinformation. Engaging with ethicists, social scientists, and the broader community in these discussions is crucial for foresight and responsible development.

Finally, the very notion of
ownership and control over the AI systems that are co-created is an ethical consideration that warrants

careful thought. Who ultimately owns the AI? Who has the authority to modify or deploy it? In participatory models, there is a compelling argument for shared ownership or at least a significant degree of shared governance. This is particularly relevant for AI systems developed by or for specific communities. For instance, an AI developed by a local indigenous community to preserve their language and cultural heritage should ideally be owned and controlled by that community. Establishing clear agreements about intellectual property, data governance, and decision-making authority is essential to ensure that the AI serves the interests of its contributors and the community it is intended to benefit. This can involve establishing community trusts, cooperative ownership models, or robust governance frameworks that empower participants with meaningful control over the AI's trajectory. By embracing these ethical considerations, participatory AI development can move beyond a transactional exchange of data and design for a more reciprocal, respectful, and ultimately beneficial relationship between AI creators and the people they serve.

The development of AI systems, particularly in fields as sensitive and impactful as healthcare, presents a unique landscape for the application of participatory design. This section delves into a case study, offering a practical illustration of how embracing collaborative principles can lead to the creation of AI-powered healthcare solutions that are not only technologically sound but also deeply aligned with the needs, values, and realities of their end-users. This hypothetical project, which we will refer to as the "MediPal AI Initiative," aimed to create an AI-driven personal health assistant designed to support individuals managing chronic conditions, such as diabetes and heart disease, at home. The journey of MediPal AI serves as a testament to the transformative power of user involvement, illuminating the tangible benefits and critical lessons learned throughout its development lifecycle.

The genesis of the MediPal AI Initiative lay in the recognized gap between existing healthcare support tools and the lived experiences of patients managing complex chronic illnesses. While many digital health solutions existed, they often lacked the personalized touch, the intuitive interface, and the holistic understanding of patient needs that are crucial for sustained engagement and effective self-management. The project team, comprising AI engineers, UX designers, and healthcare professionals, understood from the outset that a top-down approach would likely result in a product that missed the mark. Therefore, a core tenet of the MediPal AI Initiative was the commitment to a deeply participatory design process, engaging a diverse group of stakeholders from the earliest stages of ideation.

The initial phase of the MediPal AI Initiative focused on deep immersion and understanding. Instead of directly soliciting feature requests, the team initiated a series of ethnographic studies and in-depth interviews with three key groups: patients actively managing chronic conditions, their primary caregivers, and the healthcare professionals who regularly treated them. For the patients, this involved understanding their daily routines, their challenges in tracking symptoms, medication adherence, dietary needs, and the emotional toll of living with a chronic illness. For caregivers, the focus was on their role in supporting the patient, the burden of care, and their own informational needs. For clinicians, it was about understanding their workflow, the information they found most critical from patient self-monitoring, and the barriers to effective patient engagement outside of clinical settings. These sessions were not merely about gathering data; they were about building empathy and establishing a foundation of trust.

A significant challenge emerged during these early engagements: the diversity of technological literacy and comfort levels among the target user groups. Some patients were highly proficient with smartphones and digital applications, while others struggled with even basic digital interactions. Similarly, caregivers and clinicians represented a broad spectrum of tech-savviness. This realization immediately informed the design direction for MediPal AI. A crucial outcome of this early participatory phase was the decision to prioritize an exceptionally intuitive and accessible user interface. This meant opting for clear, large fonts, straightforward navigation, voice command capabilities, and a visual design that was calming and non-intimidating. The team recognized that for MediPal AI to be effective, it needed to be usable by everyone, regardless of their digital background.

Following the initial immersion, the project moved into an iterative design and prototyping phase, a critical juncture where participatory principles were applied with heightened intensity. Instead of presenting finished prototypes, the team adopted an "artifact-based" co-design approach. They started with low-fidelity prototypes – sketches, wireframes, and even paper mock-ups – and brought them to focus groups composed of patients, caregivers, and clinicians. These early sessions were invaluable for identifying fundamental usability issues and conceptual misunderstandings. For instance, early wireframes depicting medication reminders were often misinterpreted by patients as direct orders rather than suggestions, leading to anxiety. This feedback prompted a redesign that introduced more nuanced language and options for user-configurable reminder intensity and types.

As the prototypes evolved into more interactive digital mock-ups, the participatory sessions became more focused on specific interaction flows and features. For example, when designing the symptom tracking module, different patient groups offered distinct perspectives. Some preferred a simple, quick check-in, while others desired the ability to input detailed notes and qualitative descriptions of their symptoms. The team facilitated a deliberative process where users discussed these differing needs, ultimately leading to a flexible system that allowed for both quick logging and detailed entries, with clear visual indicators to differentiate between the two. Clinicians also provided critical input here, advising on the most clinically relevant data points to collect and how they should be presented in a summary format for easy review during appointments. This collaborative approach ensured that the data collected by MediPal AI was not only comprehensive but also clinically meaningful.

Data collection and the ethical considerations surrounding it were central to the MediPal AI Initiative. Recognizing the sensitivity of health data, the team prioritized transparency and user control from the outset. During the participatory sessions, the team explicitly discussed the types of data MediPal AI would collect (e.g., blood glucose readings, blood pressure, medication intake, reported symptoms, activity levels) and
why this data was necessary for the AI to provide personalized support and insights. This was not a one-time conversation; consent was revisited at multiple points in the development process. For example, when introducing the concept of the AI learning from user patterns to offer predictive insights (e.g., "You often report increased fatigue after a period of low activity"), the team ensured that participants understood this was an opt-in feature, with clear explanations of how their data would be anonymized and aggregated for pattern recognition.

A particularly impactful aspect of the participatory design was the development of the AI's conversational interface. The goal was to create an AI that was empathetic, supportive, and easy to communicate with. This involved extensive role-playing exercises and natural language processing (NLP) refinement sessions with the user groups. Patients provided examples of how they would naturally talk about their health concerns, their frustrations, and their successes. Caregivers offered insights into the language that would be most reassuring and helpful for their loved ones. Clinicians helped to ensure that the AI's responses were medically accurate and avoided giving advice that could contradict professional medical guidance. This iterative dialogue allowed the AI's language model to be trained on a rich, contextually relevant dataset, moving beyond generic responses to a more personalized and nuanced conversational style. For instance, instead of a generic "How are you feeling?", the AI learned to ask, "How has your energy level been today, and have you noticed any specific triggers for your fatigue?"

The challenges encountered were as illuminating as the successes. One significant hurdle was managing the diverse expectations and sometimes conflicting priorities of the different user groups. For example, while patients prioritized ease of use and emotional support, clinicians were primarily focused on the accuracy and clinical utility of the data. Caregivers, meanwhile, often emphasized practical alerts and the reduction of their own caregiving burden. The participatory design process provided a structured forum for these differing needs to be aired, debated, and ultimately, where possible, integrated. This often involved difficult design trade-offs, where the team had to explain the rationale behind certain decisions and seek consensus. For instance, a feature that allowed detailed symptom journaling, highly desired by some patients, was simplified in its primary view to avoid overwhelming less tech-savvy users, with an "advanced details" option readily available for those who wanted it.

Another challenge was ensuring equitable participation. The project team actively worked to recruit participants from various socioeconomic backgrounds, age groups, and ethnicities, recognizing that chronic conditions disproportionately affect certain communities. They provided stipends for participants' time, offered flexible meeting schedules, and conducted sessions in accessible community locations. For participants with visual or hearing impairments, assistive technologies were integrated into the prototypes, and feedback on their usability was actively sought. This commitment to inclusivity was not just an ethical imperative but also a practical necessity for ensuring that MediPal AI would be a truly universal solution.

The success of the MediPal AI Initiative was measured not just by its technical sophistication but by its real-world impact. Post-launch evaluations, which continued to incorporate user feedback through in-app surveys and follow-up interviews, indicated high levels of user satisfaction. Patients reported feeling more empowered in managing their health, better connected to their care team, and less alone in their journey. Clinicians found the summarized patient data provided by MediPal AI to be a valuable adjunct to their consultations, leading to more informed discussions and potentially better health outcomes. Caregivers expressed relief at having a reliable tool to support their loved ones, reducing their own stress and workload.

The lessons learned from the MediPal AI Initiative are profound. Firstly, it underscored the indispensable value of user empathy and deep understanding. AI development in healthcare cannot succeed by solely focusing on algorithms and data points; it must be grounded in the human experience of illness and care. Secondly, it highlighted the power of iterative co-creation. By involving users at every stage, from conceptualization to refinement, the team was able to course-correct early and often, preventing costly and time-consuming design errors. Thirdly, it reinforced the critical importance of transparency and ethical rigor, particularly concerning health data. Users must be active participants in deciding how their most sensitive information is used.

Moreover, the case of MediPal AI demonstrates that participatory design is not a mere add-on; it is a fundamental shift in how AI for healthcare is conceived and built. It moves from a paradigm of "designing for users" to "designing with users," fostering a sense of ownership and agency among those who will ultimately rely on the technology. The collaborative approach, while demanding and resource-intensive, yielded an AI solution that was not only effective but also ethically sound, user-centric, and truly supportive of individuals navigating the complexities of chronic health conditions. This case study serves as a compelling argument for embedding participatory principles into the very fabric of AI development in healthcare, promising a future where technology genuinely serves humanity's well-being.

Chapter 4: Understanding Machine Learning for Designers

The world of Machine Learning (ML) can often sound like a labyrinth of complex algorithms and arcane jargon, particularly for those whose expertise lies in human-centered design rather than computational science. However, understanding the fundamental principles of ML is becoming increasingly crucial for designers aiming to integrate intelligent features into their products and services. This section aims to demystify these core concepts, translating the abstract into the tangible through clear explanations, relatable analogies, and practical examples. The goal is not to turn designers into ML engineers, but to equip them with a solid conceptual grasp of what ML is, how it works at a high level, and what its inherent capabilities and limitations are. This foundational knowledge will empower more effective collaboration with AI specialists and foster more innovative and informed design decisions when conceptualizing AI-driven functionalities.

At its heart, Machine Learning is a subfield of Artificial Intelligence that enables systems to learn from data without being explicitly programmed for every possible scenario. Instead of engineers writing rigid sets of rules for every eventuality, ML models are trained on vast amounts of data, allowing them to identify patterns, make predictions, and derive insights. Think of it like teaching a child to recognize a cat. You don't list every single characteristic of every cat that has ever existed. Instead, you show the child numerous examples of cats – different breeds, colors, sizes, and poses. Over time, the child begins to understand what features are common to cats and can then identify a new cat they've never seen before. ML models learn in a fundamentally similar way.

There are several primary ways in which ML systems learn, each suited to different types of problems. The most common paradigms are supervised learning, unsupervised learning, and reinforcement learning. Understanding these distinctions is key to grasping the diverse applications of ML.

Supervised learning is perhaps the most intuitive and widely used form of machine learning. In this approach, the algorithm is trained on a labeled dataset. This means that for every data point, there is a corresponding "correct answer" or "label." The model's task is to learn the mapping between the input data and its correct output. Imagine you are teaching a student to identify different types of fruit. You would show them a picture of an apple and say, "This is an apple." Then, you show them a banana and say, "This is a banana." You continue this process with many examples, each time providing the correct label. The student (the ML model) learns to associate the visual features of each fruit with its name.

In a practical ML context, this could involve training a model to classify emails as "spam" or "not spam." The training data would consist of thousands of emails, each pre-labeled by humans as either spam or not spam. The ML model analyzes the content, sender, subject line, and other features of these emails. It learns to identify patterns that are characteristic of spam (e.g., certain keywords, unusual sender addresses, excessive punctuation) and patterns that are typical of legitimate emails. Once trained, the model can then be presented with a new, unlabeled email and predict with a certain degree of accuracy whether it is spam or not.

Another common application of supervised learning is regression, where the goal is to predict a continuous numerical value rather than a discrete category. For example, if you wanted to predict the price of a house, you would train a supervised learning model on a dataset of houses that includes features like square footage, number of bedrooms, location, and age, along with their actual sale prices. The model would learn the relationship between these features and the price. When you input the details of a new house, it can predict its likely selling price. This is what powers many real estate websites that offer estimated home values.

The effectiveness of supervised learning hinges entirely on the quality and quantity of the labeled data. If the labels are incorrect, or if the dataset is not representative of the real-world scenarios the model will encounter, the model will learn flawed patterns and perform poorly. This is why data annotation and curation are such critical steps in supervised ML projects. For designers, understanding this means recognizing that the "intelligence" of a supervised ML feature is directly tied to the data it was trained on, and any biases or inaccuracies in that data will likely be reflected in the system's behavior.

Unsupervised learning, on the other hand, deals with unlabeled data. Here, the ML algorithm is given a dataset and tasked with finding patterns, structures, or relationships within it on its own, without any predefined "correct answers." It's like giving that child a pile of assorted toys and asking them to sort them into groups. They might group them by color, by size, by type (cars, dolls, blocks), or by some other criteria they discover. They are finding inherent structure in the data without being told what that structure should be.

One of the most common tasks in unsupervised learning is clustering. Clustering algorithms group similar data points together. Imagine you have a large dataset of customer purchase histories. An unsupervised learning algorithm could be used to identify distinct customer segments based on their buying habits. It might discover a group of customers who frequently buy electronics, another group who are avid readers, and a third group who are primarily interested in home goods. These segments are not defined beforehand; the algorithm discovers them by analyzing the data. This is incredibly valuable for targeted marketing, personalization, and understanding user behavior.

Another application of unsupervised learning is dimensionality reduction. In complex datasets with many features, it can be challenging to visualize or analyze the data effectively. Dimensionality reduction techniques aim to reduce the number of features while retaining as much of the important information as possible. This can help to simplify models, speed up training, and improve performance by removing redundant or noisy features. Think of it as summarizing a lengthy book into its key plot points – you lose some detail, but you retain the essence of the story.

Anomaly detection is another powerful use case for unsupervised learning. By understanding what is "normal" within a dataset, unsupervised models can identify outliers or unusual data points that deviate significantly from the norm. This is crucial for fraud detection, identifying faulty equipment, or spotting unusual network activity. For example, in credit card transactions, an unsupervised model can learn typical spending patterns for a user and flag transactions that are significantly different, potentially indicating fraud.

The key difference and challenge with unsupervised learning is that since there are no labels, it can be harder to evaluate the "correctness" of the patterns discovered. The interpretation of these patterns often requires human expertise and domain knowledge. For designers, this means that unsupervised learning can be a powerful tool for discovery and insight generation, but the insights themselves will need careful validation and interpretation.

Reinforcement learning (RL) takes a different approach altogether, inspired by behavioral psychology. In RL, an agent learns to make a sequence of decisions by interacting with an environment. The agent receives rewards for making good decisions and penalties for making bad ones. Its goal is to learn a strategy, or "policy," that maximizes its cumulative reward over time.

Think of teaching a dog to perform a trick. When the dog performs the desired action, you give it a treat (a reward). If it does something else, you might withhold the treat or give a mild correction (a penalty, though in many RL systems, it's simply the absence of a reward). Through trial and error, the dog learns which actions lead to treats.

In ML, the "agent" could be a robot learning to navigate a maze, a game-playing AI learning to win at chess, or an algorithm optimizing traffic light timings. The "environment" is the context in which the agent operates (the maze, the chessboard, the city's traffic system). The "actions" are the choices the agent can make (move left, move piece, change light color). The "rewards" are the signals that tell the agent if it's doing well or poorly.

A classic example is training an AI to play a video game. The agent controls the game's character. It can move, jump, shoot, etc. The environment is the game world. The reward could be points scored or progress made towards a goal. The penalty could be losing a life or failing a level. Through millions of attempts, the RL agent learns which sequences of actions lead to the highest scores and eventual victory.

Reinforcement learning is particularly well-suited for problems that involve sequential decision-making and optimization over time. This includes applications like robotics, autonomous driving, resource management, and game playing. It's also the technology behind some of the most impressive AI achievements, such as AlphaGo's victory over the world's best Go players.

One of the challenges with reinforcement learning is that it can require a vast number of interactions with the environment to learn effectively, which can be computationally expensive and time-consuming, especially in real-world scenarios. Furthermore, defining an appropriate reward function can be tricky; if the reward is not aligned with the true objective, the agent might learn to exploit the system in unintended ways. For designers, understanding RL means recognizing its potential for optimizing complex processes and enabling adaptive behaviors, but also its need for careful definition of goals and environments.

Beyond these three core paradigms, there are other important concepts to touch upon. Deep Learning, for instance, is a subset of ML that uses artificial neural networks with multiple layers (hence "deep"). These deep neural networks are capable of learning complex patterns from raw data, often without requiring extensive manual feature engineering. They are behind many of the recent breakthroughs in areas like image recognition, natural language processing, and speech synthesis. Think of a deep neural network as a sophisticated system of interconnected "neurons" that process information in stages, with each layer learning to recognize increasingly abstract features. An early layer might detect edges and corners in an image, while deeper layers might combine these to recognize shapes, then objects, and eventually entire scenes.

Feature engineering is another critical concept, especially in traditional ML. This refers to the process of selecting, transforming, and creating features from raw data that best represent the underlying problem to the ML algorithm, leading to improved model accuracy and performance. For example, in predicting customer churn, raw data might include customer demographics, purchase history, and website interaction logs. Feature engineering might involve creating new features like "average purchase value," "frequency of login," or "time since last purchase." While deep learning often reduces the need for manual feature engineering by learning features directly from data, understanding what constitutes a useful feature remains vital for effective ML design.

Model training and evaluation are the processes by which ML models are built and assessed. Training involves feeding the data to the algorithm and allowing it to learn the patterns. Evaluation is the process of testing the trained model on unseen data to determine its performance, accuracy, and generalization capabilities. This often involves metrics like precision, recall, F1-score, and accuracy, depending on the type of ML task. For designers, understanding that models are not perfect and require rigorous testing is crucial for setting realistic expectations about the performance of AI-driven features.

Bias in ML is a significant ethical concern and a practical challenge. ML models learn from data, and if that data reflects historical or societal biases, the model will inevitably learn and perpetuate those biases. This can lead to discriminatory outcomes, such as facial recognition systems that perform poorly on certain demographic groups or hiring algorithms that disadvantage particular candidates. Recognizing the potential for bias in data and models is a vital part of responsible AI design. Designers and engineers must be vigilant in identifying and mitigating these biases through careful data selection, model design, and ongoing monitoring.

Explainability and interpretability are also growing areas of importance. As ML models become more complex, especially deep learning models, understanding *why* a model makes a particular prediction or decision can become difficult. This "black box" problem is a barrier to trust and adoption, particularly in high-stakes domains like healthcare or finance. Research into explainable AI (XAI) aims to develop techniques that make ML models more transparent and their decisions understandable to humans. For designers, this means considering how an AI system's reasoning can be communicated to users, fostering trust and enabling them to understand and, if necessary, contest the AI's outputs.

Finally, the concept of data drift is important for understanding the long-term performance of ML models. Data drift occurs when the statistical properties of the data on which a model was trained change over time in the real world. For instance, consumer preferences might shift, or new trends might emerge, rendering the original training data less representative. This can lead to a degradation in model performance. Therefore, ML models often need to be retrained periodically with updated data to maintain their accuracy and relevance. This iterative lifecycle of data collection, model training, deployment, monitoring, and retraining is fundamental to the ongoing success of any ML-powered system.

By demystifying these core concepts – supervised learning, unsupervised learning, reinforcement learning, deep learning, feature engineering, training, evaluation, bias, explainability, and data drift – designers can move beyond viewing machine learning as an inscrutable black box. Instead, they can begin to see it as a powerful set of tools and methodologies that, when understood conceptually, can be leveraged to create innovative, effective, and responsible user experiences. This foundational understanding is the first step towards meaningful collaboration with AI engineers and towards designing the next generation of intelligent products and services.

Types of Machine Learning Models

Having grasped the fundamental paradigms of how machine learning systems learn – supervised, unsupervised, and reinforcement learning – we can now delve into the specific types of models that implement these learning approaches. These models are the workhorses of AI, each designed to tackle particular kinds of problems. For designers, understanding these model types is crucial for identifying opportunities where AI can solve user needs and for collaborating effectively with AI engineers to select the most appropriate technology for a given task. We will explore some of the most prevalent categories: regression, classification, clustering, and the increasingly influential realm of deep learning.

Regression models are primarily used in supervised learning and are designed to predict a continuous numerical value. Imagine you're trying to forecast something that can exist on a spectrum, rather than a distinct category. This could be anything from the temperature tomorrow, the price of a stock, the lifespan of a piece of equipment, or the expected revenue for a quarter. The core idea is to establish a relationship between one or more input variables (features) and a continuous output variable. For instance, a real estate company might use regression to predict the selling price of a house. The input features could include the size of the house in square feet, the number of bedrooms, its location (perhaps represented by a numerical score based on neighborhood desirability or proximity to amenities), the age of the property, and recent comparable sales in the area. The regression model, trained on historical data of houses that have been sold, learns the underlying patterns and correlations between these features and their actual sale prices. When a new house is put on the market, its features can be fed into the trained model, which will then output a predicted selling price. This allows sellers to set realistic expectations, buyers to make informed offers, and real estate agents to provide data-driven advice.

Another common application of regression is in forecasting. Businesses often need to predict future demand for their products or services to manage inventory, staffing, and marketing efforts effectively. A retail company might use regression to predict daily sales of a particular item based on factors like the day of the week, seasonality, promotional activities, competitor pricing, and even external factors like local weather conditions or major events. Similarly, financial institutions might employ regression models to predict interest rate fluctuations or the expected credit risk of a loan applicant. The output of a regression model is a single, continuous number, and the model's performance is typically evaluated by how close its predictions are to the actual observed values, often measured by metrics like Mean Squared Error (MSE) or Root Mean Squared Error (RMSE). For designers, regression models are key to building features that involve forecasting, estimating, or quantifying outcomes that are not discrete. Think of a personalized fitness app that predicts a user's likely weight loss over a month based on their current activity levels, dietary habits, and historical progress, or a smart home system that predicts energy consumption for the next hour to optimize appliance usage.

Classification models, also a cornerstone of supervised learning, are used when the goal is to assign data points to predefined categories or classes. Unlike regression, where the output is a number, classification outputs a label. This is fundamental to many decision-making processes and is prevalent in everyday technology. A prime example is spam detection in email. When an email arrives, a classification model analyzes its content, sender, subject line, and other characteristics to decide whether it belongs to the "spam" class or the "not spam" class. The model is trained on a large dataset of emails that have been meticulously labeled by humans as either spam or not spam. It learns to identify patterns – certain keywords, the presence of attachments, unusual sender domains, excessive punctuation – that are indicative of spam. Once trained, it can then classify new, incoming emails with a high degree of accuracy.

Beyond spam, classification models are used extensively for image recognition. Consider how your smartphone camera can automatically detect faces in a photo, identify different objects, or even categorize your pictures by content (e.g., "people," "pets," "food," "landscapes"). This is powered by classification models trained on vast datasets of labeled images. Similarly, in the medical field, classification models can assist in diagnosing diseases by analyzing medical images like X-rays or MRIs, or by examining patient symptoms and medical history to predict the likelihood of a particular condition. Natural language processing (NLP) also relies heavily on classification. Sentiment analysis, for instance, uses classification models to determine if a piece of text expresses positive, negative, or neutral sentiment. This is invaluable for understanding customer feedback from reviews, social media posts, or support tickets.

Another important class of classification is binary classification, which deals with only two possible outcomes (e.g., yes/no, true/false, spam/not spam).

Multiclass classification extends this to situations where there are more than two categories. For example, a model used to identify different types of fruits in an image would be a multiclass classifier, distinguishing between apples, bananas, oranges, and so on. The performance of classification models is typically measured using metrics like accuracy, precision, recall, and the F1-score, which help to understand how well the model distinguishes between classes, particularly in scenarios where the classes might be imbalanced (e.g., detecting rare diseases). For designers, classification models enable features that make decisions, categorize information, or identify specific entities. Examples include an e-commerce platform that automatically tags products based on their descriptions, a content moderation system that flags inappropriate comments, or a banking app that categorizes transactions for budgeting purposes.

Clustering models operate within the realm of unsupervised learning. Their primary purpose is to discover hidden patterns and structures in unlabeled data by grouping similar data points together. Unlike regression and classification, where the "correct answer" is provided during training, clustering algorithms find inherent groupings without prior knowledge of what those groups should be. This makes them incredibly powerful for exploratory data analysis and for uncovering insights that might not be immediately obvious.

A quintessential application of clustering is customer segmentation. Businesses often have vast amounts of data about their customers – purchase history, browsing behavior, demographics, engagement with marketing campaigns, etc. – but this data might not be inherently organized. Clustering algorithms can analyze this data and identify distinct groups of customers who share similar characteristics. For example, a retail company might discover segments such as "high-spending loyal customers," "occasional discount shoppers," "newly acquired customers," or "price-sensitive bargain hunters." Understanding these segments allows businesses to tailor their marketing strategies, product offerings, and customer service approaches to better meet the specific needs and preferences of each group. This leads to more effective campaigns and improved customer satisfaction.

Beyond customer segmentation, clustering is used in a variety of other domains. In document analysis, it can group similar articles or news stories together, making it easier to digest large volumes of information or to identify trending topics. In biology, clustering can be applied to gene expression data to identify groups of genes that behave similarly, potentially indicating related functions. In image analysis, clustering can be used for image segmentation, grouping pixels that belong to the same object or region within an image. It's also employed in anomaly detection; by identifying what constitutes a "normal" cluster of data points, outliers that don't fit neatly into any cluster can be flagged as potential anomalies, which is crucial for fraud detection, network intrusion detection, or identifying manufacturing defects. The challenge with clustering is that the interpretation and validation of the discovered clusters often require human expertise and domain knowledge, as there are no predefined labels to verify against. For designers, clustering models offer a way to uncover underlying user typologies, identify emergent patterns in user behavior, or group content meaningfully. This can lead to features that personalize experiences based on discovered user segments or organize information more intuitively.

Deep Learning represents a significant evolution and a powerful subset of machine learning, particularly within the domain of neural networks. While traditional machine learning often requires extensive manual feature engineering – the process of selecting and transforming raw data into features that algorithms can understand – deep learning models, with their multiple layers of interconnected nodes (hence "deep"), can learn hierarchical representations of data directly from raw inputs. This capability has revolutionized fields like computer vision and natural language processing.

At the heart of deep learning are artificial neural networks, inspired loosely by the structure and function of the human brain. These networks consist of layers of "neurons" that process information. The input layer receives the raw data. This data then passes through one or more "hidden layers," where complex computations and transformations occur. Each neuron in a hidden layer takes inputs from the previous layer, applies a mathematical function (an activation function), and passes its output to the next layer. The "depth" refers to the number of these hidden layers; deeper networks can learn more complex and abstract features. Finally, an output layer produces the model's prediction or classification.

One of the most prominent applications of deep learning is in image and video analysis. Deep convolutional neural networks (CNNs) are particularly adept at processing grid-like data such as images. They can automatically learn to detect features ranging from simple edges and corners in early layers to complex object parts and entire objects in deeper layers. This has led to remarkable advances in tasks like image classification (e.g., identifying cats, dogs, cars), object detection (locating specific objects within an image), and image segmentation (delineating the boundaries of objects). Think of the facial recognition technology on your smartphone, the automated systems that identify products in online shopping images, or the sophisticated perception systems in autonomous vehicles – all heavily rely on deep learning for visual understanding.

Natural Language Processing (NLP) has also been transformed by deep learning, especially with the advent of transformer architectures. Recurrent Neural Networks (RNNs) and Long Short-Term Memory

(LSTM) networks were early successes in handling sequential data like text, but transformers have proven even more powerful in capturing long-range dependencies in language. This has enabled significant progress in tasks such as machine translation (e.g., Google Translate), text summarization, question answering, and generating human-like text. Large Language Models (LLMs) like those that power advanced chatbots and writing assistants are prime examples of deep learning's impact on language understanding and generation. For designers, these capabilities translate into features like sophisticated voice assistants that can understand complex commands, translation tools that enable seamless cross-cultural communication, content generation tools that help draft emails or reports, and systems that can analyze and summarize vast amounts of textual information.

Speech recognition is another area where deep learning excels. Systems that convert spoken language into text, like those used in virtual assistants (Siri, Alexa, Google Assistant) or dictation software, are powered by deep learning models trained on massive datasets of human speech. These models learn to map acoustic signals to phonemes, words, and sentences, handling variations in accents, background noise, and speech speed.

While deep learning offers immense power, it also comes with its own set of considerations. Deep learning models typically require enormous amounts of labeled data for training, which can be costly and time-consuming to acquire. They are also computationally intensive, requiring specialized hardware like GPUs (Graphics Processing Units) for efficient training and inference. Furthermore, deep learning models are often considered "black boxes" due to their complexity, making it challenging to understand exactly *why* they make a particular decision. This lack of interpretability can be a barrier in applications where transparency and accountability are critical, such as in healthcare or finance. Researchers are actively developing techniques in Explainable AI (XAI) to make these models more transparent.

For designers, deep learning models represent the cutting edge of AI capabilities, enabling features that can perceive, understand, and generate human-like content. When considering deep learning, it's important to think about the data requirements, computational resources, and the need for interpretability in the context of the user experience. It's a powerful tool for tackling complex, unstructured data problems that were previously intractable.

In summary, the diverse landscape of machine learning models offers a rich toolkit for designers. Regression models address problems of prediction and estimation on a continuous scale. Classification models are ideal for categorizing data into distinct groups. Clustering models uncover hidden structures and segments within unlabeled data. And deep learning models, with their layered architectures, push the boundaries of AI in understanding complex data like images, text, and speech. Each of these model types has a unique set of strengths and is suited to different kinds of design challenges, forming the bedrock upon which intelligent products and services are built. Understanding these categories provides designers with the vocabulary and conceptual framework to identify AI opportunities and to engage in informed discussions with technical teams about the most effective ways to bring AI-driven innovation to life.

Data is the lifeblood of machine learning. Without it, even the most sophisticated algorithms are merely theoretical constructs, incapable of learning, adapting, or making predictions. For designers venturing into the realm of AI, understanding the nature of this fuel – its characteristics, its journey, and its impact – is paramount. It's not enough to simply know that models need data; designers must understand *what kind* of data, *how* it's prepared, and *why* its quality and composition are inextricably linked to the success, fairness, and usability of the AI systems they help create.

The journey of data into a machine learning model is often referred to as the "data pipeline." This isn't a single, monolithic process, but rather a series of interconnected stages, each crucial for transforming raw information into a format that algorithms can comprehend and learn from. At the very beginning of this pipeline lies
data collection. This is where the raw ingredients are gathered. The sources can be incredibly diverse,

ranging from user interactions on a website or app (clicks, searches, form submissions), sensor readings from devices (temperature, location, motion), publicly available datasets (government census data, scientific research), or even data generated through human annotation. The choice of data collection methods directly impacts the scope and potential bias of the eventual model. For instance, if a sentiment analysis model is trained only on product reviews from a single platform, it might not accurately capture the nuances of sentiment expressed on social media or in forum discussions. Designers play a role here by envisioning the user interactions or data-generating events that could provide valuable insights, and by considering ethical implications of data capture from the outset. They must ask: what user behaviors or system states do we need to observe to achieve our goal? And how can we capture this information in a way that respects user privacy and builds trust?

Following collection, the data often enters a stage of **data cleaning and preprocessing**. Raw data is rarely pristine. It's common to encounter missing values (e.g., a user didn't provide their age), inconsistent formats (e.g., dates entered as "MM/DD/YYYY" and "DD-MM-YY"), or erroneous entries (e.g., an age of 200 years). The cleaning process involves identifying and rectifying these issues. This might mean imputing missing values with statistically derived estimates, standardizing formats, or removing outright errors. For example, if a model is predicting house prices, and some entries for "square footage" are clearly nonsensical (like 0 or a massive, unrealistic number), these need to be corrected or removed. Inconsistent date formats could lead to misinterpretations of temporal patterns. Similarly, text data might need to be cleaned of special characters, HTML tags, or irrelevant punctuation before it can be processed. The goal of cleaning is to create a tidy, consistent dataset that accurately reflects the phenomenon being studied. Designers might not be directly writing the code for data cleaning, but they need to understand its importance. They can anticipate potential data quality issues based on the user experience they are designing. For example, if a form allows free-text input, designers should consider how to guide users towards providing cleaner, more structured data, or how to design the system to handle variations gracefully during the cleaning phase. The effectiveness of the cleaning process directly impacts the reliability of the insights derived from the data. A model trained on dirty data is like a chef trying to cook with spoiled ingredients – the outcome will likely be poor, regardless of the chef's skill.

Once the data is cleaned, it often moves into the realm of
feature engineering. This is where raw data is transformed into meaningful "features" – the specific attributes or variables that the machine learning model will use to learn and make predictions. This is a creative and critical step, often requiring domain expertise and a deep understanding of the problem. For instance, in predicting customer churn (whether a customer will stop using a service), raw data might include individual transaction records, website login times, and customer support interactions. Feature engineering would

involve creating new features that capture patterns from this raw data. This could include: "average monthly spending," "frequency of support calls in the last quarter," "time since last purchase," or "number of active subscriptions." The creation of these features transforms abstract data points into concrete, interpretable signals that a model can latch onto. A designer's intuition about user behavior can be invaluable here. For example, a designer might notice that customers who haven't engaged with a new feature within a week of its release are more likely to churn. This observation could lead to the creation of a feature like "time since exposure to new feature" or "engagement with new feature within X days." Similarly, for a recommendation system, raw user interaction data (clicks, views) might be engineered into features like "number of times user viewed items in category X," "average rating of items previously purchased," or "time spent browsing category Y." The process of feature engineering is iterative; designers and engineers may work together to hypothesize about which engineered features are most likely to be predictive of the desired outcome. It's a testament to the idea that AI is not just about algorithms, but also about insightful data representation. The quality of engineered features can often have a more significant impact on model performance than the choice of the algorithm itself.

In many machine learning tasks, particularly in supervised learning, data needs to be **labeled**. Labeling means attaching the "correct answer" to each data point. For a spam detection model, this means each email in the training set is marked as either "spam" or "not spam." For an image recognition model, each image of a cat is labeled "cat." For a regression task predicting house prices, each historical house listing must have its actual sale price attached. Labeling is often a laborious and expensive process, typically requiring human annotators. The accuracy of these labels is paramount; if the labels are incorrect, the model will learn the wrong patterns. This is where the concept of **data quality** becomes critically important, extending beyond mere cleanliness to encompass the accuracy and consistency of the ground truth itself. Imagine training a medical diagnostic AI: if a significant number of X-rays showing a particular condition are incorrectly labeled as healthy, the model will fail to detect that condition, potentially with severe consequences. Designers can contribute by understanding how the labeling process will work and by designing user interfaces or workflows that make manual labeling as efficient and accurate as possible. For instance, a well-designed annotation tool can significantly reduce errors and speed up the process. Furthermore, designers must consider the potential for human bias to creep into the labeling process. Different annotators might interpret ambiguous cases differently, leading to inconsistencies. Establishing clear annotation guidelines and performing inter-annotator agreement checks are crucial steps to mitigate this.

The concept of
data quantity is also fundamental. Machine learning models, especially complex ones like deep neural networks, are data-hungry. The more high-quality,

relevant data a model is trained on, the better it is likely to generalize and perform accurately on unseen data. A model trained on a few hundred examples might struggle to capture the full diversity of patterns present in the real world, leading to overfitting – where the model performs well on the training data but poorly on new data. Conversely, a model trained on millions of diverse examples is more likely to learn robust representations. For designers, this means understanding that the feasibility of an AI feature often depends on the availability of sufficient data. If a novel AI capability is envisioned, but there's no readily available or collectible dataset of the required size and quality, the project might need to be re-scoped or rely on alternative approaches. Designers might need to think about how to incentivize data generation through user interactions, or how to leverage existing, albeit smaller, datasets in conjunction with techniques like transfer learning (where a model trained on a large dataset for one task is adapted for a different but related task).

Beyond quantity,
data representativeness is perhaps the most critical aspect, especially concerning fairness and equity in AI. A dataset is considered representative if it accurately reflects the diversity of the population or the environment in which the AI system will operate. If the training data is skewed, the model will inherit that skew, leading to biased outcomes. For example, if a facial recognition system is trained predominantly on images of people with lighter skin tones, it is likely to perform poorly and inaccurately when trying to

recognize individuals with darker skin tones. This is not because the algorithm is inherently racist, but because it has not been adequately exposed to the full spectrum of human diversity during its learning phase. This can have serious repercussions, from misidentification by law enforcement to discriminatory access to services.

Designers have a crucial role in identifying and mitigating these representational biases. They must advocate for data collection strategies that ensure diversity across relevant dimensions – race, gender, age, socioeconomic status, geographic location, dialect, and so on. This might involve actively seeking out data from underrepresented groups or ensuring that sampling methods are designed to capture this diversity. When working with existing datasets, designers should push for audits that reveal potential imbalances. If a dataset is found to be unrepresentative, designers can work with data scientists to explore strategies for augmentation (creating synthetic data for underrepresented groups) or re-weighting (giving more importance to data from underrepresented groups during training), always with an eye toward ethical implications and potential unintended consequences. The goal is to build AI systems that work equitably for everyone, not just for the majority represented in the training data. This requires a proactive and critical approach to data sourcing and evaluation throughout the design and development process.

Consider the development of an AI-powered hiring tool. If the historical hiring data used for training primarily reflects a workforce that is predominantly male in certain technical roles, the AI might learn to favor male candidates, even if implicitly. This bias could manifest in how resumes are scored or how candidate profiles are ranked. A designer, understanding the importance of diversity and inclusion, would question the composition of the training data. They might ask: "Does this data reflect the diversity we aim for in our workforce?" If not, they would collaborate with HR and data science teams to source more balanced data, perhaps by including data from successful candidates from diverse backgrounds or by using techniques to "debias" the existing data. They might also design the user interface of the hiring tool to highlight potential biases to the human reviewers, providing them with information to make more equitable decisions. This moves beyond simply building a functional tool to building a responsible and ethical one.

Another area where data is central is in the development of **personalization** features. To personalize a user experience - whether it's recommending products, tailoring news feeds, or adjusting difficulty levels in a game - the system needs data about the individual user. This data might include their past interactions, stated preferences, demographic information, or even inferred interests. The quality, quantity, and representativeness of this user-specific data are vital. If the system only has a limited history of a user's interactions, its personalization efforts might be generic or even irrelevant. If the data collected about a user is inaccurate or based on flawed assumptions, the personalization will be misguided. For designers, this means thoughtfully considering what user data is necessary and sufficient for effective personalization, and designing clear, transparent mechanisms for users to understand and control the data being used. Features like "Why am I seeing this?" or options to explicitly manage preferences are crucial for building user trust when personalization relies on data. Moreover, designers must be mindful of the potential for "filter bubbles" or echo chambers that can arise from hyper-personalization if the data doesn't account for introducing users to diverse perspectives or content.

The design of AI systems is, therefore, deeply intertwined with the data pipeline. Designers are not just creating interfaces; they are, in essence, designing the conditions under which AI models learn. This involves a continuous cycle of understanding user needs, identifying the relevant data required to meet those needs, considering how that data will be collected, cleaned, engineered, and labeled, and critically evaluating the quantity and representativeness of the data to ensure performance, fairness, and ethical integrity. Ignoring the data aspect is akin to a chef focusing solely on presentation without considering the quality of the ingredients – the final product will inevitably be compromised.

Looking ahead, as AI systems become more integrated into our lives, the responsibility of designers to champion robust, ethical, and representative data practices will only grow. This requires a willingness to engage with the technical intricacies of the data pipeline, to ask challenging questions, and to collaborate closely with data scientists and engineers. It means understanding that every piece of data collected, every feature engineered, and every label assigned is a design choice with tangible consequences for the AI system's behavior and its impact on users and society. Data is not just the fuel; it is the blueprint, the foundation, and the ongoing nourishment for intelligent systems. Designers who master this understanding are equipped to build AI that is not only intelligent but also responsible, equitable, and truly beneficial. They move from simply "using AI" to actively "designing with AI" at its most fundamental level.

In the intricate dance between human interaction and artificial intelligence, a crucial step bridges the chasm between raw, unrefined information and the sophisticated logic of machine learning models: feature engineering. This is where the art of design meets the science of data, transforming observable user behaviors, contextual nuances, and intrinsic attributes into the very language that algorithms understand. It's not merely about feeding data into a system; it's about intelligently curating and crafting that data into potent signals that illuminate patterns and drive predictive power. For designers, grasping the essence of feature engineering is akin to becoming a master storyteller for the machine, articulating the user's narrative through a series of precisely defined variables.

Imagine a smart home system aiming to optimize

energy consumption. The raw data might include temperature readings from various sensors, occupancy status in rooms, time of day, and even external weather forecasts. While individually these points are informative, their true predictive potential is unlocked through feature engineering. A designer, with an understanding of user habits and environmental factors, might propose creating features such as: "average daily temperature fluctuation," "percentage of unoccupied rooms during peak hours," "time elapsed since last manual thermostat adjustment," or "predicted solar gain based on weather and window orientation." Each of these engineered features condenses complex, multi-faceted information into a concise, digestible input for the machine learning model. The "average daily temperature fluctuation," for instance, captures the system's tendency to experience rapid changes, which might signal an inefficient heating or cooling setup or a user who frequently adjusts the thermostat. The "percentage of unoccupied rooms during peak hours" directly speaks to potential energy waste. By translating raw sensor data and user actions into these meaningful features, designers enable the AI to learn more nuanced patterns and make more intelligent, energy-saving decisions. This process requires empathy for the user and a deep understanding of the physical environment, allowing designers to anticipate what aspects of the user's life and environment are most relevant to the AI's task.

The power of feature engineering lies in its ability to amplify the insights derived from user context. Consider a mobile application designed to promote healthy habits. Raw data might include GPS location, accelerometer readings, screen-on time, and app usage logs. A designer might identify that a user's physical activity is a key predictor of their adherence to a fitness regimen. However, simply knowing the raw distance traveled might not be sufficient. Through feature engineering, one could derive features like: "average daily step count," "duration of continuous walking or running sessions," "frequency of transitions from sedentary to active states," or "deviation from usual activity patterns on weekends." These engineered features provide a richer, more context-aware representation of the user's physical activity. For example, "duration of continuous walking" might differentiate between a short walk to the car and a sustained period of exercise, which could be a more direct indicator of health-conscious behavior. Similarly, detecting "transitions from sedentary to active states" could highlight spontaneous bursts of activity, important for understanding overall movement patterns. The designer's understanding of what constitutes "healthy activity" directly informs the creation of these features, making the AI more adept at recognizing and encouraging such behaviors. This moves beyond a simple count of steps to a qualitative understanding of the user's movement.

In e-commerce, personalization is paramount, and feature engineering plays a vital role in understanding customer intent and preferences. Raw data might consist of product views, items added to cart, purchase history, search queries, and time spent on product pages. A designer working on a recommendation engine would seek to transform this into features that predict future purchasing behavior. This could involve creating features like: "number of distinct categories browsed in the last week," "average price point of items viewed vs. items purchased," "time since last purchase in a specific category," "recency and frequency of searches for particular keywords," or "ratio of items added to cart vs. items ultimately purchased." The feature "average price point of items viewed vs. items purchased" can reveal whether a user is a bargain hunter, a brand loyalist, or someone exploring high-end options. The "ratio of items added to cart vs. items ultimately purchased" could indicate indecisiveness or price sensitivity, guiding the recommendation strategy. By crafting these features, the designer helps the AI understand not just *what* a user looked at, but *how* they interacted with it and what their underlying intentions might be, leading to more relevant and persuasive recommendations.

Moreover, feature engineering allows designers to imbue AI models with an understanding of temporal dynamics and sequential patterns, which are often critical for predicting user behavior over time. For a streaming service, raw data might include viewing history, search logs, and subscription duration. To predict churn (a user cancelling their subscription), features could be engineered to capture changes in viewing habits. Examples include: "trend in viewing hours over the last month," "decrease in engagement with new releases," "frequency of browsing but not playing content," or "increase in the time gap between consecutive viewing sessions." A declining trend in viewing hours, especially in conjunction with a lack of engagement with new content, might be a strong precursor to churn. A designer can identify these patterns by observing user behavior and translate them into robust features that signal potential disengagement. This foresight allows the AI to proactively address potential issues, perhaps by offering personalized content suggestions or special promotions to retain the user. Without these engineered temporal features, the AI might only react to a cancellation after it has occurred, rather than intervening preemptively.

The concept of creating interaction-based features is also immensely valuable. For a customer support chatbot, raw data might include transcripts of user queries, the time taken to resolve issues, and customer satisfaction ratings. Engineered features could describe the nature and complexity of user problems. For instance: "number of distinct topics covered in a single conversation," "frequency of keywords indicating frustration (e.g., 'confused,' 'difficult,' 'error')," "average duration of user pauses during interaction," or "proportion of queries requiring escalation to a human agent." A high frequency of "frustration keywords" or long "user pauses" might indicate a poorly designed user flow or a complex problem that the chatbot is struggling to address. This information can then be used not only to improve the chatbot's responses but also to flag potentially dissatisfied customers for follow-up. The designer, by understanding the user's journey through the support process, can hypothesize about which interaction patterns are indicative of success or failure, and thus create the features that best capture these dynamics.

Feature engineering is not solely about capturing user actions; it's also about translating static or semi-static user attributes into actionable data points. Consider a financial planning application. Raw data might include age, income, occupation, and stated financial goals. While these are useful, engineered features can provide a more dynamic and insightful view. Examples could be: "debt-to-income ratio," "savings rate as a percentage of income," "estimated risk tolerance based on stated goals and age," or "time horizon to achieve stated goals." A low savings rate combined with ambitious goals, for instance, highlights a potential mismatch that the AI can address with tailored advice. The "estimated risk tolerance" can be a composite feature derived from multiple inputs, allowing the AI to suggest investment strategies that align with the user's comfort level. Designers, by understanding the principles of financial health and user psychology, can guide the creation of these synthesized features, making the AI's advice more pertinent and effective.

The creative aspect of feature engineering allows designers to go beyond obvious metrics and capture more subtle aspects of user experience. For an educational platform, raw data might include quiz scores, time spent on lessons, and video completion rates. A designer might create features that gauge user engagement and learning progression in a more holistic way: "rate of improvement in quiz scores over consecutive modules," "proportion of optional learning materials accessed," "diversity of topics explored beyond assigned curriculum," or "pattern of engagement with collaborative features (if any)." A high "rate of improvement" is a strong indicator of effective learning, while "diversity of topics explored" suggests genuine intellectual curiosity. These features move beyond simple performance metrics to capture deeper aspects of the learning journey, enabling the AI to identify students who are not just passing but truly thriving and those who might need more personalized support.

Furthermore, feature engineering enables the incorporation of external or contextual data that might not be directly generated by user interaction but is highly relevant to the AI's task. For an app that tracks and predicts traffic conditions, raw data might include vehicle speed, GPS locations, and road sensor data. Engineered features could incorporate real-time weather information ("current precipitation level," "visibility reduction due to fog"), local event schedules ("presence of major sporting events or concerts nearby"), or even historical traffic patterns for specific times and days of the week ("average congestion level at this intersection on a Tuesday afternoon"). By merging these external data streams with user-generated data, the AI can achieve far more accurate predictions. A designer can identify which external factors are most likely to influence user outcomes and advocate for their inclusion in the feature set. For example, a designer might realize that a sudden downpour can drastically alter commute times, even if vehicle speeds haven't yet reflected this change, and ensure that weather data is a prominent feature.

The process of feature engineering is inherently iterative and collaborative, often requiring close partnership between designers and data scientists. Designers bring an understanding of user needs, goals, and contexts, while data scientists possess the technical expertise to implement and validate features. A designer might hypothesize that a user's "hesitation time" before making a purchase is a strong indicator of price sensitivity. They would articulate this hypothesis to a data scientist, who would then figure out how to operationalize it – perhaps by measuring the time between adding an item to the cart and proceeding to checkout, or the time spent hovering over the "buy" button. Together, they would experiment, build, and refine these features, observing their impact on the AI's performance. This collaborative loop is essential for ensuring that the engineered features are not only technically sound but also meaningfully aligned with the human experience the AI is meant to serve.

It's also important to recognize that feature engineering can sometimes involve a degree of abstraction and simplification. For example, if an AI is designed to understand user sentiment from text, raw data might be individual words or character sequences. Feature engineering could involve techniques like: "counting the frequency of positive and negative words from a predefined lexicon," "identifying the presence of emoticons or punctuation that indicate emotion," or "calculating the average sentence length in a review." While complex natural language processing models can sometimes bypass explicit feature engineering by learning representations directly from raw text, simpler models or those requiring interpretability often benefit greatly from these handcrafted features. The designer's role here is to understand what aspects of the text are most indicative of sentiment and to guide the creation of features that capture these.

The ethical implications of feature engineering are also significant. When creating features, designers and data scientists must be mindful of potential biases that can be inadvertently encoded. For instance, if a feature is derived from data that is unrepresentative of certain demographic groups, the AI will inherit that bias. A designer might propose a feature related to "creditworthiness" based on factors like income and employment history. However, if the historical data used to define "good creditworthiness" disproportionately reflects affluent populations, this feature could unfairly disadvantage individuals from lower socioeconomic backgrounds. Therefore, a critical part of feature engineering involves auditing the proposed features for fairness and equity, ensuring that they do not perpetuate or amplify societal biases. Designers must advocate for features that are predictive and fair, working to identify and mitigate potential harms before they are embedded in the AI system. This might involve exploring alternative features or techniques to de-bias existing ones.

In essence, feature engineering is the creative engine that allows designers to translate the rich, complex tapestry of human context into the structured, quantifiable language of machine learning. It is the process by which a designer's understanding of user needs, behaviors, and environments becomes directly embedded within the AI's learning mechanism. By thoughtfully crafting features that capture essential signals – whether it's a user's activity patterns, their shopping habits, their learning progression, or even their emotional state – designers empower AI systems to be more accurate, more personalized, and more relevant. This is not merely a technical step in the data pipeline; it is a deeply design-driven activity that shapes the very intelligence and behavior of the AI, making it a powerful tool for creating truly user-centered and impactful experiences. The ability to foresee what data points will be most telling and how to represent them effectively is a hallmark of an AI-savvy designer.

The journey of a machine learning model from a nascent idea to a functional, intelligent system is characterized by a critical phase: **model training and evaluation**. This isn't a single, definitive step, but rather a dynamic, iterative process that breathes life into algorithms, transforming them from inert code into capable decision-makers. For designers, understanding this core process is paramount, as it directly dictates the efficacy, reliability, and ultimately, the user-centricity of the AI solutions they help shape.

At its heart, model training is akin to teaching a student. You provide the student with a vast amount of information – the training data – and guide them to identify patterns, relationships, and underlying principles within that information. In the context of machine learning, this "information" is the meticulously engineered data we discussed previously, rich with features that represent user behaviors, contextual nuances, and environmental factors. The "student" is the ML algorithm, a mathematical framework designed to learn from data. Through a series of complex computations, the algorithm adjusts its internal parameters, essentially tuning itself to recognize specific patterns within the training data. The goal is to build a model that can generalize, meaning it can apply what it has learned from the training data to new, unseen data it has never encountered before. This generalization is the true test of a well-trained model; it signifies that the model has moved beyond memorizing specific examples and has developed a true understanding of the underlying task.

Imagine training a model to predict whether a user will click on an advertisement. The training data would consist of numerous past instances, each labeled with whether a click occurred or not. Features might include user demographics, browsing history, the content of the advertisement, the time of day, and the device used.

During training, the algorithm would sift through these thousands or millions of data points, gradually learning which combinations of features are most strongly associated with a click. For instance, it might discover that users who have previously shown interest in a particular product category are more likely to click on ads related to that category, especially during specific hours. The training process iteratively refines the model's internal "weights" or "coefficients" associated with each feature, making it progressively better at distinguishing between a click and a no-click scenario. The more data, and the more relevant the features, the more refined and accurate this learning process can become.

However, simply completing the training process doesn't guarantee a successful model. A model can become overly specialized, memorizing the training data so perfectly that it fails to perform well when presented with slightly different, real-world data. This phenomenon is known as "overfitting." Conversely, a model might fail to learn enough from the training data, resulting in poor performance on both seen and unseen data; this is called "underfitting." This is where **evaluation** becomes indispensable. Evaluation is the rigorous assessment of a trained model's performance, using data that was **not** used during the training phase. This "unseen data," often referred to as a validation set or a test set, provides an unbiased measure of how well the model is likely to perform in the real world.

Think of the student again. After studying for an exam,

they wouldn't just review their notes; they would take a practice test. This practice test simulates the actual exam, revealing areas where their understanding is solid and where it needs further improvement. Similarly, evaluating an ML model involves presenting it with the validation or test data and comparing its predictions to the actual outcomes. This comparison allows us to quantify the model's performance using various
evaluation metrics. These metrics are crucial for understanding not just *if* the model is working, but *how well* it is working, and in what specific ways it might be succeeding or falling short.

Several key evaluation metrics are commonly employed, each offering a different perspective on model performance.
Accuracy, perhaps the most intuitive, measures the overall proportion of correct predictions made by the model. If a model predicts a click correctly 90% of the time across the test set, its accuracy is 90%. While useful, accuracy can be misleading, especially in situations where the data is imbalanced - meaning one outcome is far more frequent than another. For instance, in predicting rare diseases, a model that simply predicts "no disease" for every patient might achieve high accuracy if the disease is uncommon, but it would be practically useless.

This is where metrics like
precision and **recall** become vital, particularly in classification tasks where the goal is to identify specific instances. **Precision** answers the question: "Of all the

instances the model predicted as positive, how many were actually positive?" High precision means that when the model flags something as positive (e.g., predicts a user will click), it is highly likely to be correct. This is crucial when the cost of a false positive is high. For example, in a spam detection system, high precision ensures that legitimate emails are not incorrectly flagged as spam.

Recall, on the other hand, answers: "Of all the actual positive instances, how many did the model correctly identify?" High recall means the model is good at finding all the positive cases. This is essential when the cost of a false negative is high. For instance, in a medical diagnosis scenario for a life-threatening illness, high recall is paramount to ensure that as many actual cases as possible are identified, even if it means a few false positives are flagged for further investigation.

The relationship between precision and recall is often a trade-off. Improving one can sometimes lead to a decrease in the other. Designers and data scientists must work together to determine which metric is more critical for a given application, based on the user experience and the potential consequences of errors. For example, in a content recommendation system, a designer might prioritize recall to ensure users are exposed to a wide variety of potentially interesting content, accepting that some recommendations might not be perfect. Conversely, in a financial fraud detection system, precision would be paramount to minimize the disruption caused by incorrectly flagging legitimate transactions.

Other important metrics include the **F1-score**, which provides a harmonic mean of precision and recall, offering a single score that balances both. The **Area Under the Receiver Operating Characteristic Curve (AUC-ROC)** is another powerful tool that assesses a model's ability to distinguish between classes across various probability thresholds. It's particularly useful for understanding how well a model performs when faced with different levels of sensitivity and specificity.

For designers, the overarching takeaway from the model training and evaluation process is its **iterative nature**. Building a high-performing AI model is rarely a linear path. It involves a continuous cycle of:

1. **Data preparation and feature engineering:** Crafting the right inputs.
2. **Model selection:** Choosing an appropriate algorithm.
3. **Training:** Teaching the model from the data.
4. **Evaluation:** Assessing performance with unseen data.
5. **Analysis and iteration:** Identifying weaknesses and returning to earlier steps to refine features, adjust model parameters, or even select a different model.

This cyclical approach underscores the importance of establishing clear **performance goals and benchmarks** from the outset. Before diving into training, designers, in collaboration with stakeholders and data scientists, must define what constitutes success. What level of accuracy is acceptable for this application? What is the maximum acceptable rate of false positives or false negatives? These benchmarks should be directly tied to **user-centric objectives**. For instance, if the goal is to reduce customer support wait times, a benchmark might be a model that can accurately categorize incoming queries with 95% precision, enabling efficient routing. If the aim is to enhance user engagement on a streaming platform, a benchmark might be a recommendation model that achieves a certain click-through rate on suggested content.

Without well-defined goals, evaluation becomes a nebulous exercise, making it difficult to determine if the model is truly meeting user needs. It's not enough for a model to simply "learn" patterns; it must learn patterns that translate into desirable user experiences and reliable real-world performance. This requires designers to think critically about the implications of model performance metrics on the end-user. A model with high accuracy might still lead to a frustrating user experience if its errors occur in critical decision points or disproportionately affect certain user segments.

Furthermore, evaluation is not a one-time event that concludes with model deployment. As real-world data evolves and user behaviors shift, a model's performance can degrade over time. This necessitates **continuous monitoring and re-evaluation** of deployed models to ensure they remain effective and aligned with their intended purpose. This ongoing oversight is crucial for maintaining user trust and the overall integrity of the AI system. By actively participating in and understanding the model training and evaluation phases, designers can ensure that the AI systems they help create are not just technically sound, but also robust, reliable, and truly serve the needs of the humans they are designed to assist. This deep understanding empowers them to ask the right questions, challenge assumptions, and ultimately, champion the development of AI that is both intelligent and ethically grounded.

Chapter 5: Auditing AI for Bias and Fairness

The journey from raw data to intelligent decision-making within an AI system is fraught with potential pitfalls, and one of the most insidious is the infiltration of algorithmic bias. While we strive to build systems that are objective and fair, the reality is that AI models can, and often do, perpetuate and even amplify existing societal inequalities. This isn't typically a result of malicious intent on the part of the developers; rather, it's an unintended consequence woven into the fabric of the data used for training and the assumptions embedded in the design process itself. Understanding how this bias creeps in is the first crucial step towards identifying and mitigating it through rigorous auditing.

At its root, algorithmic bias often stems from the data itself. AI models learn by identifying patterns within vast datasets. If these datasets reflect historical or societal biases, the model will inevitably learn and reproduce those biases. Consider, for instance, historical hiring data. If, for decades, a particular industry or role has predominantly hired individuals from a certain demographic group (due to past discriminatory practices, not necessarily merit), a model trained on this data might learn to associate that demographic with success in that role, even if the underlying skills or qualifications are equally present in other groups. This creates a self-perpetuating cycle where the AI continues to favor the historically dominant group, making it harder for underrepresented individuals to break in, irrespective of their abilities. This isn't because the algorithm is intentionally prejudiced, but because it is a faithful, albeit uncritical, reflection of the biased world it has been shown. The data, in this scenario, acts as a mirror, but one that reflects a distorted, inequitable reality.

The "unintended consequence" aspect of bias is further amplified by the very nature of machine learning. These systems excel at correlation, not necessarily causation or fairness. They identify statistical associations within the data. If, for example, in a dataset of past loan applications, certain zip codes or neighborhoods are disproportionately associated with loan defaults, an algorithm might learn to associate those geographical markers with higher risk. This could lead to individuals from those areas being unfairly denied loans, not because of their individual creditworthiness, but because of a statistical correlation tied to their location. This correlation may itself be a symptom of systemic issues, such as historical redlining or economic disparities, rather than a direct indicator of an individual's ability to repay. The AI, in its quest to optimize for predicted risk, inadvertently entrenches these socio-economic disadvantages.

Flawed assumptions in model design can also be a significant source of bias. When engineers and data scientists build an AI system, they make decisions about what data to collect, what features to engineer, and what objective functions to optimize. These decisions are often informed by their own perspectives, which can, consciously or unconsciously, carry their own biases. For example, when designing a facial recognition system, if the training data overwhelmingly comprises images of individuals with lighter skin tones, the system may perform poorly when trying to identify individuals with darker skin tones. This isn't necessarily because the designers actively disliked darker skin; it might be due to a lack of diverse data collection, an assumption that the primary use cases would involve lighter-skinned individuals, or an oversight in ensuring equitable representation. The resulting system, therefore, becomes less effective and potentially more intrusive for certain groups, simply because the design process didn't adequately account for the full spectrum of human diversity.

Another critical area where bias can manifest is in the definition and measurement of "success" for the AI. If an AI system is optimized solely for a narrow definition of efficiency or profit, it may overlook fairness as a crucial outcome. In the context of hiring algorithms, for example, an AI might be designed to identify candidates who are most likely to be "good hires" based on historical patterns of employee retention and performance. However, if those historical patterns are themselves a product of biased hiring practices, the AI will simply learn to replicate those biased patterns. It will prioritize candidates who fit the mold of previously successful hires, effectively filtering out potentially excellent candidates who don't fit the established, and potentially biased, profile. The pursuit of a quantifiable metric like "retention rate" can, in this instance, mask and perpetuate underlying discrimination.

The impact of algorithmic bias is not theoretical; it has tangible and often devastating real-world consequences. One of the most widely discussed examples is the bias found in facial recognition systems. Studies have repeatedly shown that these systems exhibit significantly lower accuracy rates for women and individuals with darker skin tones compared to white men. This disparity can lead to serious repercussions, from wrongful arrests due to misidentification to the denial of access to services or opportunities. Imagine an individual being flagged by a security system based on a faulty facial recognition match, leading to suspicion, interrogation, or even arrest. The cost of such an error, for the individual, is immense, and it disproportionately affects already marginalized communities. The technology, designed for security and efficiency, becomes an instrument of injustice.

Similarly, biased algorithms have been observed in the realm of criminal justice. Risk assessment tools used to predict recidivism (the likelihood of a defendant re-offending) have been found to disproportionately assign higher risk scores to Black defendants compared to white defendants, even when controlling for similar criminal histories and other relevant factors. This can influence critical decisions such as bail amounts, sentencing, and parole. The consequence is that individuals from certain racial groups are more likely to be detained pretrial, receive harsher sentences, or be denied parole, not necessarily because they are inherently more dangerous, but because the algorithm reflects and amplifies systemic biases present in the data it was trained on. This can trap individuals in a cycle of incarceration, perpetuating racial disparities within the justice system.

The financial sector also provides stark examples of algorithmic bias. Loan application systems, credit scoring models, and even insurance underwriting algorithms have been shown to discriminate against certain groups. For instance, studies have indicated that algorithms used to approve or deny loans can inadvertently penalize applicants from lower-income neighborhoods or those who do not fit traditional employment patterns, even if they possess the financial means to repay. This can create barriers to economic mobility, making it harder for individuals and communities already facing financial challenges to access essential services and build wealth. The promise of objective, data-driven financial decision-making can, in practice, reinforce existing economic inequalities, creating a less equitable playing field for everyone.

Even seemingly innocuous applications can harbor bias with significant downstream effects. Recommendation engines, for example, which suggest content, products, or services, can exhibit bias by reinforcing existing preferences and limiting exposure to diverse viewpoints. If an algorithm primarily recommends content aligned with a user's past consumption habits, it may inadvertently create an "echo chamber," shielding the user from alternative perspectives and reinforcing stereotypes. For instance, a news recommendation system might, based on past clicks, consistently present articles that align with a particular political ideology, thereby solidifying that viewpoint and reducing the likelihood of engagement with opposing arguments. While not as immediately severe as a wrongful arrest, this can contribute to societal polarization and a diminished capacity for informed discourse.

The insidious nature of algorithmic bias lies in its often-invisible presence. Unlike overt human prejudice, algorithmic bias can be deeply embedded within complex systems, making it difficult to detect without deliberate auditing. The logic of the algorithm may appear sound, its statistical performance high, yet its outcomes can be profoundly unfair. This is why the concept of "auditing AI for bias and fairness" is not merely an academic exercise but a critical necessity for responsible AI development and deployment. It's about peeling back the layers of complex computation to examine the underlying data, the design choices, and the ultimate impact of these intelligent systems on individuals and society. Without this critical examination, we risk building a future where technological advancement inadvertently entrenches and amplifies the very inequalities we seek to overcome. The pervasiveness of this bias demands a proactive and systematic approach to ensure that AI serves humanity equitably and justly.

Bias in artificial intelligence is not a singular, monolithic problem but rather a multifaceted issue that can arise from various stages of an AI system's lifecycle. To effectively audit AI for fairness, it is imperative to understand these distinct points of origin. These sources can broadly be categorized into three interconnected areas: biased datasets, algorithmic bias inherent in the model's architecture and learning process, and interaction bias, which emerges from the dynamic relationship between the AI and its users over time. Each of these categories represents a critical junction where inequities can be introduced or amplified, necessitating careful scrutiny.

The most frequently cited and perhaps the most pervasive source of bias stems from the data used to train AI models. Datasets are not neutral repositories of facts; they are often a reflection of the world as it is, with all its historical baggage and inherent inequalities. Historical inequities are deeply embedded in the data. For example, if historical hiring records show a consistent underrepresentation of women in leadership roles, an AI trained on this data might learn to associate male candidates with leadership potential more strongly, not because of innate capability, but because of the historical patterns presented. This is not an indictment of the data scientists who gather this information, but rather an acknowledgment that the past often carries forward its prejudices. Similarly, in the criminal justice system, if policing has historically been more concentrated in certain neighborhoods or targeted specific demographic groups, arrest data will reflect this, leading to AI systems that may unfairly flag individuals from those communities as higher risk, irrespective of their individual actions. The data, in this instance, becomes a perpetuator of past discriminatory practices, casting a long shadow over future decisions.

Beyond historical biases, sampling issues within datasets can also introduce significant skew. Datasets are rarely perfect representations of the entire population or phenomenon they are meant to model. If a dataset is not collected in a way that ensures diversity and representativeness across all relevant subgroups, the resulting AI model will inevitably be biased. For instance, if a facial recognition system is trained primarily on images of individuals with lighter skin tones, it will likely perform poorly when attempting to identify individuals with darker skin tones. This is not due to any inherent flaw in the concept of facial recognition itself, but because the training data failed to adequately capture the full spectrum of human appearance. This can lead to disparities in accuracy, where the technology is less reliable and potentially more intrusive for certain populations. Another example could be in medical diagnostic AI. If a dataset for training a skin cancer detection algorithm predominantly features images of lighter skin, the model might be less effective at identifying melanomas on darker skin, leading to delayed or missed diagnoses for a significant portion of the population. The error here lies in the unrepresentative sampling, which leads to a system that is less equitable in its performance.

The very process by which AI models learn can also be a source of bias, often referred to as algorithmic bias. Machine learning algorithms are designed to find patterns and optimize for specific objectives. However, the way these patterns are identified and the objectives are defined can inadvertently create or amplify bias, even if the training data itself were perfectly balanced and representative. One way this occurs is through feature selection and engineering. Data scientists choose which characteristics (features) of the data the AI should pay attention to. If the chosen features are proxies for protected attributes (like race or gender), even indirectly, the algorithm can learn to discriminate. For example, using a postal code as a feature in a loan application model might seem innocuous, but if certain postal codes are heavily correlated with specific racial or socioeconomic groups due to historical housing patterns, the algorithm could effectively discriminate based on race or class without explicitly being told to do so. The algorithm is not inherently prejudiced; it is simply following the statistical correlations it has been shown, but those correlations are themselves tainted by societal biases.

Moreover, the optimization process itself can lead to biased outcomes. AI models are typically trained to minimize a loss function, which quantifies the error of the model's predictions. If the loss function is not carefully designed to account for fairness, the model might prioritize overall accuracy at the expense of fairness for minority groups. For instance, in a fraud detection system, a model might achieve very high overall accuracy by correctly identifying most non-fraudulent transactions and a reasonable proportion of fraudulent ones. However, if the errors disproportionately affect a particular demographic group (e.g., flagging legitimate transactions from that group as fraudulent more often), the system, while technically "accurate" by a broad metric, is unfair. This is sometimes referred to as "disparate impact," where a seemingly neutral policy or algorithm has a disproportionately negative effect on a protected group. The algorithm has learned to be good at its primary task, but this efficiency has come at the cost of equitable treatment.

The complexity of deep learning models also makes them susceptible to generating emergent biases. These "black box" models can develop internal representations and decision-making processes that are not easily interpretable. It is possible that during the intricate process of training, the model latches onto spurious correlations or develops internal logic that, while mathematically sound from its perspective, leads to discriminatory outcomes when applied to the real world. The sheer number of parameters and the non-linear relationships within deep neural networks can obscure how bias is being introduced, making it harder to pinpoint the exact source within the model's architecture or training process. Without specialized auditing techniques, these hidden biases can persist and go undetected.

A third critical area where bias can be introduced or exacerbated is through interaction bias. This refers to how users interact with the AI system over time, and how these interactions, in turn, influence the AI's behavior and future decisions. Many AI systems, particularly those involving machine learning, are not static; they learn and adapt based on new data and feedback. If users interact with the system in biased ways, or if the system is deployed in environments where biased human decisions are prevalent, the AI can begin to learn and reflect these user-driven biases.

Consider recommendation systems, for example. If a streaming service's recommendation algorithm learns from user viewing habits, and if a particular user consistently chooses to watch content that perpetuates stereotypes, the algorithm will learn to recommend more of that content. Over time, the user's "personalized" experience becomes an echo chamber that reinforces their existing biases. This is particularly problematic when these recommendation systems are used for more critical applications, such as job recommendations or educational content. If the system learns to steer certain demographics towards specific types of jobs or learning paths based on biased user interactions or initial data, it can limit opportunities and create a self-fulfilling prophecy of limited career or educational trajectories.

Another manifestation of interaction bias occurs when

AI systems are used to augment human decision-making. In such scenarios, the human operator might develop an over-reliance on the AI's recommendations, especially if the AI is perceived as objective or infallible. If the AI itself contains subtle biases, the human operator might not critically evaluate its output, inadvertently accepting and acting upon biased suggestions. Conversely, if the human operator has their own biases, they might selectively heed the AI's recommendations when they align with their own preconceived notions, and dismiss them when they don't, thereby reinforcing their own biases through the "validation" of the AI. This creates a feedback loop where human biases are amplified by the AI, and the AI's biased outputs are legitimized by human adoption. For instance, a hiring manager might use an AI tool to screen resumes. If the AI subtly favors certain candidate profiles, and the manager also has unconscious biases that align with these profiles, they might over-index on the AI's positive signals for those candidates, further marginalizing others.

Furthermore, the very act of collecting feedback from users can introduce bias if the feedback mechanisms are not carefully designed. If a system prompts users for feedback in a way that leads them to confirm existing beliefs or if only a specific subset of users provides feedback, the collected data can be skewed. This skewed feedback can then be used to retrain and update the AI, leading it to drift further away from fairness. For example, if a customer service chatbot is evaluated solely on user satisfaction ratings, and users from a particular demographic group are consistently less satisfied due to the chatbot's inherent biases (e.g., it struggles with their dialect or cultural references), the system might be updated to better serve the majority, further alienating the minority group.

Diagnosing bias at each of these stages requires a systematic approach tailored to the specific source. For biased datasets, techniques involve rigorous exploratory data analysis (EDA) to understand the distribution of features across different demographic groups. This includes looking for imbalances, missing data patterns that disproportionately affect certain groups, and correlations between features and protected attributes. Statistical tests can be employed to identify significant differences in data distributions. Visualizations, such as histograms, scatter plots, and heatmaps, are invaluable for uncovering these disparities. For instance, plotting the average salary by gender in a historical employment dataset would immediately reveal gender-based pay gaps. Similarly, examining the representation of different racial groups in image datasets used for facial recognition is crucial. Auditing tools can also be used to automatically scan datasets for known bias indicators.

When it comes to algorithmic bias, the audit focuses on the model's behavior and its decision-making processes. This involves evaluating the model's performance metrics not just overall, but disaggregated by sensitive attributes. If a loan approval model has an

80% accuracy rate overall, it's essential to check if that accuracy is consistent across different racial groups or income levels. Metrics like precision, recall, false positive rates, and false negative rates should be calculated for each subgroup. Techniques such as counterfactual fairness, which asks whether the outcome would change if a sensitive attribute were different, can be applied. Interpretability methods, such as LIME (Local Interpretable Model-agnostic Explanations) or SHAP (SHapley Additive exPlanations), can shed light on which features the model is relying on for its predictions, helping to identify if it's using proxies for protected attributes. Moreover, specific fairness-aware machine learning algorithms and regularization techniques exist that can be incorporated during training to mitigate bias.

For interaction bias, the audit becomes more longitudinal and requires analyzing logs of user interactions and system responses over time. This might involve tracking how the AI's recommendations evolve based on user engagement, or how its decision-making shifts after being exposed to specific types of user feedback. Analyzing user cohorts and their distinct interaction patterns can reveal if the system is adapting in ways that create disparate outcomes. Techniques like A/B testing can be used to compare different versions of an AI system or different interaction designs to see which leads to fairer outcomes. It's also important to consider the context in which the AI is deployed; if it's used in a setting with known human biases, the interaction patterns are likely to reflect and potentially amplify those biases. Therefore, auditing interaction bias often requires a combination of technical analysis of AI behavior and a qualitative understanding of the socio-technical environment in which the AI operates. This understanding can inform the design of more robust feedback mechanisms and user interfaces that encourage equitable interactions.

Understanding these interconnected sources of bias—data, algorithm, and interaction—is the foundational step toward building and deploying AI systems that are not only intelligent but also just. Each stage of the AI lifecycle presents an opportunity for bias to creep in, and a comprehensive audit must traverse all these potential pathways. Without this deliberate and thorough examination, the promise of AI risks being overshadowed by the reality of perpetuating and amplifying the very inequalities we seek to dismantle.

Detecting and measuring bias in artificial intelligence systems is a critical step in the auditing process. It moves beyond understanding *where* bias might originate to actively identifying and quantifying its presence and impact. This requires a systematic approach, employing both quantitative metrics and qualitative assessments to gain a comprehensive understanding of an AI's fairness. Without robust measurement, claims of fairness remain speculative, and efforts to mitigate bias lack direction.

Quantitative methods form the backbone of bias measurement, providing objective, numerical indicators of disparity. These metrics allow us to translate abstract notions of fairness into concrete, interpretable figures. One of the most fundamental concepts is **statistical parity**, also known as demographic parity. This metric assesses whether the AI system's outcomes are independent of sensitive attributes, such as race, gender, or age. In simpler terms, it asks: "Does a person from group A have the same probability of receiving a positive outcome as a person from group B, regardless of their group membership?" For example, in a loan application system, statistical parity would be achieved if the approval rate for male applicants is the same as for female applicants, or if the approval rate for applicants of different racial backgrounds is identical. Mathematically, this is often expressed as $P(\hat{Y}=1 \mid A=a) = P(\hat{Y}=1 \mid A=b)$, where $\hat{Y}$ is the predicted outcome (e.g., loan approved), and A represents the sensitive attribute with different values 'a' and 'b'. While seemingly straightforward, achieving statistical parity can sometimes lead to models that are less accurate overall or that do not reflect genuine differences in risk or propensity. It is a powerful tool for identifying disparities but should be used in conjunction with other metrics, as it does not account for the possibility that different groups might legitimately have different probabilities of a desired outcome.

Moving beyond simple outcome independence, **equalized odds** and **equal opportunity** offer more nuanced measures. These metrics focus on fairness in the classification process, particularly for binary outcomes (e.g., approved/rejected, hired/not hired, predicted positive/predicted negative). **Equalized**

odds requires that the true positive rate (sensitivity) and the false positive rate are equal across different groups. This means that for individuals who *should* receive a positive outcome (true positives), the AI correctly identifies them at the same rate across all groups. Simultaneously, for individuals who *should not* receive a positive outcome (true negatives), the AI incorrectly identifies them as positive (false positives) at the same rate across all groups. Mathematically, this translates to: $P(\hat{Y}=1 \mid Y=1, A=a) = P(\hat{Y}=1 \mid Y=1, A=b)$ (equal true positive rates) AND $P(\hat{Y}=1 \mid Y=0, A=a) = P(\hat{Y}=1 \mid Y=0, A=b)$ (equal false positive rates). Consider a medical diagnostic AI: equalized odds would imply that the AI is equally good at correctly identifying a disease in different patient populations (equal true positive rates) and equally unlikely to falsely diagnose the disease in healthy individuals across those same populations (equal false positive rates).

A related but slightly less stringent criterion is **equal opportunity**, which focuses only on the true positive rates being equal across groups, while allowing false positive rates to differ. It essentially asks: "Are individuals who are truly deserving of a positive outcome equally likely to receive it, regardless of their sensitive attribute?" Mathematically, this is $P(\hat{Y}=1 \mid Y=1, A=a) = P(\hat{Y}=1 \mid Y=1, A=b)$. This metric is particularly relevant in contexts where the cost of a false negative (failing to identify a deserving individual) is significantly higher than the cost of a false positive. For instance, in a job recruitment system, equal opportunity would ensure that qualified candidates from all backgrounds have an equal chance of being identified and advanced in the hiring process, even if the system is more prone to incorrectly flagging unqualified candidates from certain groups as potentially suitable. The choice between equalized odds and equal opportunity often depends on the specific ethical considerations and potential harms associated with different types of errors in a given application.

Disparate impact analysis, a concept originating in legal frameworks, provides another crucial lens for measuring bias. It examines whether a policy or decision-making process has a statistically significant adverse effect on individuals belonging to a particular protected group, even if the policy itself does not explicitly discriminate. This is often measured using the "four-fifths rule" (also known as the 80% rule), which suggests that a selection rate for any protected group that falls below four-fifths (or 80%) of the rate for the group with the highest rate may indicate adverse impact. For example, if a hiring algorithm has an overall selection rate of 50%, but for a particular minority group, the selection rate is only 30%, this might trigger further investigation. The calculation would be (30% / 50%) = 0.6, or 60%, which is below the 80% threshold. Disparate impact analysis is not a definitive proof of discrimination but serves as a signal that warrants closer examination of the algorithm's behavior and its underlying data. It is important to note that the four-fifths rule is a guideline, and statistical significance tests are often employed in practice to determine if observed disparities are truly anomalous.

Beyond these statistical definitions, **predictive parity** is another important metric. It requires that the precision of the model is equal across groups, meaning $P(Y=1 \mid \hat{Y}=1, A=a) = P(Y=1 \mid \hat{Y}=1, A=b)$. In other words, when the model predicts a positive outcome for an individual, the probability that the prediction is actually correct (i.e., the individual truly warrants a positive outcome) should be the same regardless of their sensitive attribute. In a credit scoring context, predictive parity would mean that if the model assigns a high credit score to an applicant, the likelihood that they are actually a good credit risk is the same, whether they are male or female, or from different ethnic backgrounds.

The challenge with these quantitative metrics lies in their inherent trade-offs. It is often impossible to satisfy all fairness criteria simultaneously. For instance, achieving perfect statistical parity might require sacrificing predictive accuracy or equalized odds. This is known as the "impossibility theorem" in algorithmic fairness, highlighting that in most real-world scenarios, different fairness definitions are mutually exclusive. Therefore, the selection of which metrics to prioritize depends heavily on the specific application, its societal context, and the potential harms associated with different types of unfairness. This necessitates a clear understanding of the ethical goals and the potential consequences of algorithmic decisions.

While quantitative metrics provide essential numerical evidence of bias, they do not tell the whole story. **Qualitative methods** play a crucial role in complementing these numbers, offering deeper insights into the lived experiences of individuals affected by AI systems and uncovering biases that might not be captured by statistical measures alone. These methods are vital for understanding the "why" behind the numbers and for identifying issues that might be overlooked by purely data-driven approaches.

One significant qualitative approach is **user studies**, which involve directly engaging with individuals who use or are impacted by the AI system. These studies can take various forms, including interviews, focus groups, and ethnographic observations. Through in-depth interviews, researchers can ask users about their perceptions of the AI system's fairness, its trustworthiness, and how it affects their decisions or opportunities. For example, in a study of a hiring AI, interviews with job applicants might reveal concerns about the system's language, its perceived biases in evaluating certain qualifications, or how it makes them feel about their chances. Focus groups allow for richer discussions, where participants can build upon each other's experiences and identify shared concerns. Ethnographic observations involve observing users interacting with the AI in their natural environments, providing context and revealing subtle ways in which bias might manifest during real-world use. For instance, observing how a customer service AI responds to different accents or dialects could highlight issues that raw accuracy metrics might miss.

Expert reviews, specifically by domain experts and ethicists, are another form of qualitative assessment. These reviews involve bringing in individuals who have deep knowledge of the AI's application domain and/or expertise in AI ethics. They can scrutinize the AI's design, its intended use, and its potential societal implications from a critical perspective. For example, an expert in criminal justice might review a risk assessment tool used in sentencing, identifying potential flaws in how it interprets factors related to socioeconomic status or past offenses, which could inadvertently perpetuate systemic biases. Ethicists can help frame fairness considerations within broader societal values and identify potential ethical dilemmas that the quantitative metrics alone might not surface. They can ask challenging questions about the distribution of benefits and harms, the potential for unintended consequences, and the alignment of the AI system with fundamental human rights.

Bias bounties and red teaming exercises can also be considered qualitative methods, though they often incorporate quantitative elements. In a bias bounty program, organizations invite external researchers and the public to find and report instances of bias in their AI systems, often in exchange for rewards. This crowdsourced approach can uncover a wide range of biases, including those that the internal team might not have anticipated. Red teaming involves intentionally trying to "break" the AI system in ways that expose its vulnerabilities, including its biases. A red team might try to trick a content moderation AI into approving hate speech, or a facial recognition system into misidentifying individuals, specifically looking for instances where these failures disproportionately affect certain demographic groups. These adversarial approaches are powerful for uncovering edge cases and systematic failures that might not be apparent during standard testing.

Furthermore, **case studies and post-deployment monitoring** provide ongoing qualitative data about bias. Examining real-world instances where an AI system has produced unfair outcomes, understanding the context, and gathering feedback from affected parties is crucial. This involves going beyond initial audits to continuously assess the AI's performance and impact in the wild. For example, if a recommendation system starts to funnel users into increasingly narrow or stereotypical content categories, qualitative feedback from users about their experience and the perceived limitations of the recommendations can be invaluable for identifying and addressing this drift.

When combining quantitative and qualitative methods, it's essential to understand their synergistic relationship. Quantitative metrics provide the "what" – they tell us *that* a disparity exists and *how large* it is. Qualitative methods, on the other hand, help us understand the "why" and the "how" – they shed light on the underlying mechanisms of bias, the user experience, and the real-world consequences. For instance, a quantitative analysis might reveal that a hiring AI has a significantly lower selection rate for female candidates in technical roles (disparate impact). A subsequent qualitative study, through interviews with female applicants and hiring managers, might uncover that the AI penalizes candidates who list collaborative projects over individual achievements, a factor more commonly emphasized in male-dominated technical cultures. This combination allows for a more complete diagnosis of the problem and informs more effective interventions.

Ultimately, the goal of bias detection and measurement is to provide a comprehensive, evidence-based understanding of an AI system's fairness. This toolkit of quantitative metrics and qualitative assessments equips designers, developers, and auditors with the means to rigorously evaluate AI systems. It moves the conversation from subjective opinion to objective analysis, ensuring that fairness is not an afterthought but a fundamental aspect of AI development and deployment. The careful selection and application of these methods are paramount to building AI that serves all members of society equitably.

Having meticulously detected and measured bias within an AI system, the subsequent and equally crucial phase is the implementation of strategies for its mitigation. This involves a proactive and systematic approach to rectify the unfair disparities uncovered during the auditing process. Bias mitigation is not a singular event but rather a continuous cycle of intervention, re-evaluation, and refinement, often necessitating a delicate balance between achieving fairness and maintaining acceptable levels of performance. The strategies employed can broadly be categorized into three stages: pre-processing, in-processing, and post-processing. Each stage offers distinct methods and considerations for debiasing AI systems, and their effectiveness can vary depending on the nature of the bias, the AI architecture, and the specific application domain.

Pre-processing techniques focus on modifying the training data before it is fed into the AI model. The rationale here is that if the data itself is biased, the model trained on it will inevitably learn and perpetuate those biases. Therefore, the goal is to create a cleaner, more equitable dataset. One common pre-processing strategy is
data re-weighting, where instances in the training data are assigned different weights. If a particular demographic group is underrepresented or has historically faced disadvantage, their data points can be given higher weights, effectively amplifying their signal during training. Conversely, overrepresented or privileged groups might have their data points down-weighted. This forces the model to pay more attention to the experiences and characteristics of the marginalized groups. For example, in a loan application dataset where historical approval rates were lower for a certain ethnic minority due to systemic discrimination, re-weighting could assign higher importance to approved applications from that group,

encouraging the model to learn the criteria associated with responsible borrowing within that community, rather than reflecting past discriminatory lending practices.

Another powerful pre-processing method is **data augmentation**, particularly relevant for domains like image recognition or natural language processing where data scarcity can exacerbate bias. Data augmentation involves creating new, synthetic data points by applying transformations to existing ones. For debiasing, these transformations can be designed to increase the representation of underrepresented groups or to neutralize biased associations. In image recognition, if a facial recognition dataset is predominantly composed of lighter skin tones, augmentation could involve applying color jittering or generating synthetic images of individuals with darker skin tones while preserving other identifying features. Similarly, in NLP, if sentiment analysis models exhibit bias against certain dialects or accents due to limited training data, augmentation can involve generating paraphrased sentences or using techniques like back-translation to increase the diversity of linguistic inputs. A more advanced form of data manipulation is **fairness-aware sampling**, where data is sampled in a way that ensures proportional representation of different groups or specific outcomes across groups. This is particularly useful when dealing with imbalanced datasets that might naturally lead to bias.

Adversarial debiasing can also be implemented at the pre-processing stage, though it's often considered an in-processing technique as well. In this approach, a secondary model, often called a "discriminator," is trained alongside the primary AI model. The discriminator's task is to predict the sensitive attribute (e.g., race, gender) from the features learned by the primary model. The primary model is then trained not only to perform its intended task (e.g., classification) but also to "fool" the discriminator, meaning it learns representations that are independent of the sensitive attribute. This adversarial game encourages the primary model to learn features that are predictive of the outcome but not predictive of the protected group membership, thereby reducing bias. For instance, if a hiring model is learning features that correlate with gender (e.g., specific extracurricular activities more common among one gender), the adversarial network would penalize the hiring model for developing such gender-specific predictive patterns.

Moving to
in-processing techniques, these methods involve modifying the AI algorithm or its training process to incorporate fairness constraints directly. The aim is to build fairness into the model's learning objective. One approach is to add **regularization terms** to the model's loss function. These terms penalize the model for exhibiting unfair behavior according to chosen fairness metrics. For example, a regularization term could be added that penalizes differences in prediction rates between different demographic groups, thus encouraging the model to converge towards a state

where outcomes are more equitable. If a model is being trained for loan approvals, and the loss function includes a penalty for a significant gap in approval rates between male and female applicants, the training process will actively try to minimize this gap alongside minimizing prediction errors.

Another in-processing strategy is **fairness-constrained optimization**. This involves explicitly defining fairness criteria as constraints within the optimization problem that the AI model solves during training. For instance, if we want to ensure equalized odds, the optimization process would aim to minimize the classification error while simultaneously ensuring that the true positive rates and false positive rates are equal across protected groups. This is mathematically formulated as an optimization problem with inequality or equality constraints that enforce fairness. This approach requires careful mathematical formulation and often sophisticated optimization algorithms, but it offers a direct way to embed fairness objectives into the model's learning.

Adversarial debiasing, as mentioned earlier, is also a prominent in-processing technique. By training a primary model to learn representations that are invariant to sensitive attributes, it directly interferes with the model's ability to leverage those attributes for prediction, thus mitigating bias. This method is particularly effective for complex models like deep neural networks, where internal representations are often opaque.

Finally,
post-processing techniques involve adjusting the model's predictions *after* they have been generated, without retraining the model or altering the training data. This is often the simplest to implement, as it doesn't require access to the training pipeline or data. The core idea is to calibrate the model's outputs to satisfy fairness criteria. One common post-processing method is **threshold adjustment**. For binary classification tasks, a decision threshold is used to convert a model's raw output score (e.g., a probability) into a final prediction (e.g., approve/reject). By adjusting these thresholds independently for different demographic groups, one can achieve fairness. For example, if a model tends to give lower scores to a particular minority group, its threshold for positive prediction might be lowered, allowing more individuals from that group to receive a favorable outcome, thereby improving demographic parity or equal opportunity. This is akin to setting different "passing grades" for different groups to achieve a desired overall distribution of outcomes.

Another post-processing technique is
reject option classification. This involves modifying the prediction for certain individuals where the model's confidence is low or where the prediction might lead to unfair outcomes. For cases that fall into a "reject region" – typically those close to the decision boundary or where fairness metrics are likely to be violated – the system can abstain from making a prediction, deferring the decision to human review. This can be particularly

useful in high-stakes applications where a potentially biased automated decision could have severe consequences. For example, if a loan application score is borderline and the model's prediction for a particular group is likely to be unfair based on historical data, the system might flag it for human underwriter review rather than making an automated decision.

It is critical to acknowledge the inherent **trade-offs between fairness and accuracy**. As noted in the context of bias measurement, achieving perfect fairness across all metrics is often mathematically impossible without sacrificing some degree of predictive accuracy. This is a fundamental challenge in AI fairness. For instance, aggressively re-weighting data to improve representation for a minority group might lead to slightly lower overall accuracy on the entire dataset, as the model has to compromise on optimizing for the majority. Similarly, adjusting thresholds in post-processing to achieve demographic parity might mean that some individuals who would have been correctly classified by the original model are now misclassified, thus reducing overall accuracy.

The decision of which trade-off to accept depends heavily on the application's context and the potential harms associated with different types of errors. In a medical diagnosis AI, a slight reduction in overall accuracy might be acceptable if it significantly improves the detection rate of a rare but serious disease in an underrepresented patient population. Conversely, in a system for predicting creditworthiness, a significant drop in accuracy could have widespread economic consequences. Ethical frameworks and stakeholder consultations are essential for navigating these trade-offs. It's not about achieving absolute fairness at any cost, but about making informed, responsible decisions about acceptable levels of fairness and accuracy for a given use case.

Furthermore, bias mitigation is rarely a one-time fix. It's an **ongoing process** that requires continuous monitoring and re-evaluation. AI systems operate in dynamic environments, and their performance can drift over time. Data distributions can change, societal norms evolve, and new forms of bias can emerge. Therefore, a system that was fair at deployment may become unfair later. This necessitates robust post-deployment monitoring systems that continuously audit the AI for bias using the metrics and qualitative methods discussed previously. When new biases are detected, mitigation strategies need to be reapplied, and the entire cycle of audit and intervention must begin anew. This iterative nature of bias mitigation underscores the importance of building flexible and adaptable AI systems.

Let's consider some **practical examples** of bias mitigation in action.

In **image recognition**, specifically facial recognition systems, bias often manifests as lower accuracy for

individuals with darker skin tones or for women. Mitigation strategies might include:
Data augmentation: Generating synthetic images of underrepresented skin tones and facial features, or applying color normalization techniques to make the model less sensitive to variations in skin pigmentation.
Re-weighting: Assigning higher weights to images of individuals from underrepresented demographic groups during training.
Fairness-aware loss functions: Incorporating terms into the loss function that penalize disparities in recognition accuracy across different racial or gender groups.
Post-processing threshold adjustment: If a facial recognition system is used for access control, its thresholds for accepting a match might be adjusted to ensure that the false rejection rate is similar across different demographic groups, even if it slightly increases the false acceptance rate for some.

In
natural language processing (NLP), bias can appear in language models, sentiment analysis, and translation systems. For instance, language models may associate certain professions with specific genders (e.g., "doctor" with male pronouns, "nurse" with female pronouns). Mitigation strategies include:
Debiasing word embeddings: Techniques like "gender-neutralizing" word vectors aim to remove gender stereotypes from word representations. For example, making the vector for "doctor" equidistant from "he" and "she."

Data balancing: Curating training datasets that contain more balanced gender associations for professions and other concepts.
Contextual debiasing: Developing models that are sensitive to the context in which words appear, allowing them to correctly infer gender-neutrality or specific associations without relying on broad stereotypes. For instance, a model should understand that "the programmer wrote code" can refer to any gender.
Adversarial debiasing: Training a language model to predict text content while simultaneously training a discriminator to predict gender from the learned representations, forcing the model to produce gender-invariant representations.

In
hiring and recruitment tools, bias can lead to unfair screening of candidates. For example, an AI might unfairly penalize resumes that use language more commonly associated with one gender, or it might favor candidates from specific universities that historically have more graduates from privileged backgrounds. Mitigation strategies include:
Feature selection and engineering: Carefully selecting features for the model that are strongly predictive of job performance and are less likely to be correlated with protected attributes. This might involve anonymizing certain fields or using proxies for skills that are not directly linked to demographic markers.
Data pre-processing: Removing potentially biased features from resumes, such as names, graduation years (which can infer age or generational biases), or specific

clubs/activities that might correlate with gender or socioeconomic status.
Fairness-aware recommendation systems: In systems that suggest candidates for open roles, ensuring that the suggestions are diverse and not overly concentrated in one demographic group.
Post-processing calibration: Adjusting the scores or ranking of candidates to ensure that qualified individuals from all backgrounds have an equitable chance of being advanced. For instance, if a candidate is highly qualified but has a slightly lower score due to a subtle bias, the score can be adjusted upwards to reflect their true potential.

In
loan and credit scoring systems, bias can result in discriminatory lending practices, denying loans to creditworthy individuals from marginalized communities. Mitigation strategies include:
Fairness metrics: Applying metrics like disparate impact, equal opportunity, and predictive parity to evaluate the fairness of the scoring model across different racial, ethnic, or gender groups.
Regularization: Adding penalties to the model's training objective that discourage discriminatory outcomes. For example, penalizing differences in predicted default rates across groups.
Counterfactual fairness: Ensuring that if an individual's sensitive attribute were different, their loan outcome would remain the same, assuming all other relevant (and non-discriminatory) factors are equal.
Threshold adjustment: As discussed, modifying the loan approval threshold for different groups to achieve

a desired level of outcome parity, carefully considering the trade-offs with accuracy and the risk of adverse selection.

The choice of mitigation strategy is not a one-size-fits-all proposition. It requires a deep understanding of the AI system, the data it uses, the specific biases identified, and the ethical goals of the application. Moreover, the effectiveness of any mitigation technique must be rigorously validated. This involves re-auditing the system after the intervention to confirm that the bias has been reduced and that the system's performance is still acceptable. This iterative process of auditing, mitigating, and re-auditing is fundamental to building responsible and equitable AI systems.

The challenge of bias mitigation also extends to the **transparency and explainability** of these processes. When mitigation techniques are applied, especially complex ones like adversarial debiasing or fairness-constrained optimization, understanding *how* they work and *why* they are effective can be difficult. This can create a "black box" effect within the debiasing process itself. Therefore, it is crucial to document the mitigation strategies employed, the rationale behind their selection, and the results of their implementation. This documentation serves not only for internal accountability but also for external audits and for building trust with stakeholders and the public.

In conclusion, strategies for mitigating AI bias form a multifaceted and evolving field. They range from data manipulation before training, to algorithmic adjustments during training, to output calibration after prediction. Each approach has its strengths, weaknesses, and associated trade-offs, particularly concerning the delicate balance between fairness and accuracy. Effective bias mitigation is not a singular solution but a continuous, iterative process that demands careful consideration of the specific context, rigorous evaluation, and a commitment to ongoing monitoring. By employing a combination of pre-processing, in-processing, and post-processing techniques, and by navigating the inherent trade-offs with ethical clarity, organizations can move towards developing AI systems that are not only intelligent but also just.

The journey through understanding and mitigating bias in AI systems culminates in a critical realization: true fairness is not a feature that can be bolted on at the end of development. Instead, it must be woven into the very fabric of the AI lifecycle, from the initial spark of an idea to the final deployment and ongoing maintenance. This section shifts our focus from reactive auditing and mitigation to a proactive, design-centric approach, underscoring the imperative of building fairness into the AI design process itself. This proactive integration is not merely an ethical nicety; it is a foundational requirement for creating AI systems that are not only effective and efficient but also trustworthy, equitable, and ultimately, beneficial to all members of society.

Establishing fairness as a core design principle from the outset demands a fundamental shift in how we conceptualize and develop AI. It means moving beyond a purely technical problem-solving mindset to one that embraces socio-technical considerations. This involves clearly articulating specific fairness goals for an AI system at its inception. These goals should be concrete, measurable, and aligned with the intended application and its potential societal impact. For instance, a hiring AI might have a goal of achieving demographic parity in shortlisting rates for qualified candidates across different racial groups. A medical diagnostic AI might aim for equal opportunity, ensuring that the true positive rates for detecting a particular disease are consistent across age and gender demographics. These initial goals serve as a compass, guiding every subsequent design and development decision.

Crucially, the process of defining these fairness goals and embedding them into the design must involve a diverse range of stakeholders. This includes not only AI developers, data scientists, and product managers but also ethicists, legal experts, domain specialists, and, most importantly, representatives of the communities who will be most affected by the AI system. Engaging diverse perspectives helps to uncover potential biases and unintended consequences that might otherwise go unnoticed. For example, a community engagement process for a predictive policing AI might reveal that certain data collection methods, while seemingly neutral, disproportionately impact marginalized neighborhoods, leading to feedback that influences data sourcing and model features. Similarly, involving users with disabilities in the design of an AI-powered accessibility tool can highlight specific accessibility requirements and potential failure modes that developers might not have anticipated. This collaborative approach fosters a deeper understanding of the problem space and ensures that the developed AI is responsive to the nuanced needs and values of its intended users and the broader public.

Beyond setting goals and engaging stakeholders, the design process itself can be augmented with specific techniques to promote fairness. This can include employing fairness-aware feature engineering, where developers actively scrutinize and select features that are less likely to be proxies for protected attributes or to carry historical societal biases. For example, instead of using a candidate's zip code, which can correlate with socioeconomic status and race, a developer might opt for features that directly measure relevant skills or experience. Data collection strategies can also be designed with fairness in mind, prioritizing diversity and representativeness from the ground up. This might involve actively seeking out data from underrepresented groups or employing data generation techniques that aim to balance existing disparities, always with careful consideration for privacy and ethical data sourcing.

Furthermore, the choice of AI architecture and algorithms can significantly impact fairness. Some models are inherently more interpretable than others, which can aid in identifying and rectifying bias. For instance, decision trees or linear models might offer greater transparency than complex deep neural networks. When deep learning models are necessary for performance, techniques for improving their interpretability, such as LIME (Local Interpretable Model-agnostic Explanations) or SHAP (SHapley Additive exPlanations), can be employed to understand why a model makes a particular decision, thereby revealing potential biased reasoning. The design process should therefore involve a conscious evaluation of these trade-offs between model complexity, performance, and the ability to ensure and verify fairness.

The concept of "responsible AI" also emphasizes the importance of transparency and explainability not just of the final model's predictions, but of the entire development process. This includes meticulously documenting every stage of design and development, from the initial problem definition and data sourcing to the model selection, training, and evaluation. Such documentation should clearly articulate the fairness goals, the methods used to achieve them, the rationale behind specific design choices, and the results of bias audits and mitigation efforts. This transparency is crucial for accountability, enabling internal review, external auditing, and building trust with the public. In high-stakes applications, the ability to explain *why* an AI system made a particular decision, especially if that decision has significant consequences for an individual, is paramount. This requires building explainability into the model architecture or employing post-hoc explanation techniques that are robust and reliable.

Establishing clear lines of accountability for fairness is another cornerstone of building fairness into the design process. In an organizational context, this means defining roles and responsibilities for ensuring AI fairness. Who is responsible for setting fairness metrics? Who oversees bias audits? Who makes the final decision when fairness and accuracy present conflicting trade-offs? Without clear accountability, fairness can become a diluted responsibility, easily overlooked in the rush to deploy a new system. This might involve establishing dedicated AI ethics committees, appointing fairness officers, or integrating fairness considerations into performance reviews for AI development teams. The goal is to create an organizational culture where fairness is not an optional add-on but an integral part of every AI project.

The development of AI systems can also benefit from the adoption of formal frameworks and guidelines for responsible AI development. Many organizations and industry bodies are producing such frameworks, which often include principles related to fairness, transparency, accountability, and robustness. Adhering to these frameworks can provide a structured approach to integrating fairness into the design process. For example, a framework might mandate specific steps for bias assessment at different stages of development, require the use of certain fairness metrics, or stipulate the level of documentation needed for AI systems. Such guidelines can serve as a blueprint, ensuring that essential fairness considerations are not overlooked.

Furthermore, the notion of "fairness by design" extends to the user interface and user experience of AI-powered applications. How an AI system presents its outputs, explains its reasoning, and allows for user feedback can significantly influence its perceived fairness and trustworthiness. For example, an AI chatbot that is designed to be transparent about its limitations and to escalate complex or sensitive queries to human agents can foster greater user trust than one that attempts to handle all situations autonomously, potentially making errors that appear arbitrary or biased. Similarly, an AI recommendation system that provides explanations for its suggestions and allows users to easily correct or refine them can be perceived as fairer than one that operates as a "black box."

The proactive integration of fairness considerations also necessitates continuous learning and adaptation within the development teams. As our understanding of AI bias evolves and new mitigation techniques emerge, design processes must be flexible enough to incorporate these advancements. This calls for ongoing training for AI developers, data scientists, and product managers on the latest research and best practices in AI fairness. It also means fostering a culture of critical self-reflection, where teams are encouraged to question their assumptions, challenge existing practices, and actively seek out opportunities to improve the fairness of the AI systems they build. This iterative learning cycle is crucial for staying ahead of the curve in the rapidly evolving landscape of AI ethics.

The commitment to building fairness into the design process also involves a deliberate effort to avoid common pitfalls. One such pitfall is the "fairness washing," where organizations pay lip service to fairness without implementing meaningful changes in their design and development practices. This can involve superficial audits or the adoption of generic fairness statements without concrete actions. Another pitfall is the tendency to over-rely on automated fairness tools without critically evaluating their underlying assumptions or the specific context of the AI application. While automated tools can be helpful, they are not a substitute for human judgment, ethical reasoning, and stakeholder engagement.

Moreover, building fairness into the design process requires a pragmatic approach to managing trade-offs. As previously discussed, achieving perfect fairness across all metrics is often impossible without sacrificing some degree of accuracy or other performance objectives. The design process should include mechanisms for systematically evaluating these trade-offs, documenting the decisions made, and ensuring that these decisions are aligned with the ethical principles and societal impact goals of the AI system. This often involves difficult conversations and a commitment to making the most responsible choices, rather than the easiest ones.

Finally, the journey towards building fairness into the AI design process is an ongoing one. It is not a one-time implementation but a continuous commitment to improvement. As AI systems are deployed and interact with the real world, new challenges and unforeseen consequences may arise. Therefore, the design process must include robust mechanisms for ongoing monitoring, evaluation, and iteration, ensuring that fairness remains a central consideration throughout the entire lifecycle of the AI system. This might involve establishing feedback loops from users and affected communities, conducting regular re-audits, and being prepared to adapt and update the system as needed. By embedding fairness considerations proactively into every stage of the AI design and development process, we can move towards creating AI systems that are not only technologically advanced but also ethically sound and truly beneficial to humanity.

Chapter 6: User Experience (UX) Design for AI

The landscape of user experience (UX) design has been irrevocably transformed by the advent of artificial intelligence (AI). Where traditional software design typically focused on predictable interfaces, deterministic functionalities, and static user flows, AI-powered systems introduce a dynamic, adaptive, and often opaque layer that necessitates a profound rethinking of how we approach user interaction. The core challenge lies in bridging the gap between the human user's expectations and the intelligent system's evolving capabilities and inherent unpredictability. This is not merely an evolution of design principles; it is a paradigm shift that demands a new set of tools, methodologies, and a fundamental understanding of what it means to design for intelligence.

Historically, UX design has been grounded in established principles of usability, learnability, efficiency, and satisfaction. We designed interfaces with clear affordances, predictable responses, and consistent navigation. A button, once clicked, would always perform the same action. A form, when filled incorrectly, would provide specific error messages. This predictability fostered user confidence and reduced cognitive load. However, AI introduces a layer of emergent behavior, learning capabilities, and autonomy that can disrupt this predictable model. An AI system is not a static set of rules; it is a learning entity that can change its behavior over time based on new data and interactions. This inherent fluidity requires a design approach that embraces, rather than resists, this dynamic nature.

One of the most significant departures from traditional UX is the need to design for systems that are not just responsive, but also adaptive and learning. Unlike a conventional application that executes pre-programmed instructions, an AI system can, and often should, learn from its users. This learning can manifest in various ways: personalizing recommendations, adjusting interface elements to suit individual preferences, or even modifying its core functionalities based on observed user behavior patterns. For example, a news aggregation AI might learn which topics a user frequently engages with and begin to prioritize those stories, subtly reshaping the user's information diet. A smart home AI might learn the user's daily routines and proactively adjust lighting, temperature, and security settings without explicit commands. The UX designer's role, therefore, extends beyond crafting static screens to designing the *learning process* itself, ensuring that the AI's adaptation is beneficial, transparent, and controllable by the user. This involves creating mechanisms for users to guide the AI's learning, provide feedback on its adaptations, and even reset its learned behaviors if they become undesirable.

This adaptive nature of AI also compels us to move beyond monolithic, one-size-fits-all interfaces. The ideal user experience in an AI-driven world is often personalized. An AI can analyze a vast array of user data – preferences, past interactions, contextual information – to tailor the interface and its functionality to the individual. Consider an AI-powered educational platform. Instead of presenting a uniform curriculum, it can dynamically adjust the difficulty, pace, and even the teaching style based on the student's learning progress, strengths, and weaknesses. The UX must therefore be designed to accommodate this personalization, ensuring that the interface can fluidly reconfigure itself to present the most relevant information and options at any given moment. This requires designing flexible UI components and interaction models that can accommodate a wide range of personalized states, rather than fixed, predetermined layouts.

However, the power of AI also brings the challenge of opacity. Many advanced AI models, particularly deep learning systems, operate as "black boxes," making it difficult to understand precisely *why* they make a particular decision or recommendation. This lack of transparency can be a significant hurdle for UX. Users are often more comfortable and trusting when they understand the rationale behind a system's actions. If an AI-powered credit scoring system denies a loan, the applicant is entitled to understand the reasons. If a medical diagnostic AI suggests a particular treatment, the patient and physician need to understand the evidence supporting that suggestion. Therefore, rethinking UX for AI necessitates the integration of explainability and transparency into the design. This doesn't always mean revealing the intricate algorithms, but rather providing clear, user-understandable explanations for the AI's outputs. This could involve presenting key factors that influenced a decision, offering confidence scores, or providing links to relevant information. The UX designer must collaborate with AI developers to surface the most crucial and understandable aspects of the AI's reasoning, transforming a black box into a more transparent and trustworthy system.

The introduction of autonomy in AI systems also poses a unique UX challenge. When an AI is empowered to make decisions or take actions without direct human intervention, the potential for unintended consequences increases. A self-driving car's AI might encounter a novel ethical dilemma, or an AI-controlled trading algorithm might execute a series of trades that destabilize a market. From a UX perspective, this means designing systems that can gracefully handle uncertainty, inform users of their autonomous actions, and provide clear pathways for human oversight and intervention. This could involve designing sophisticated alert systems, intuitive dashboards that visualize the AI's current operational state and decision-making process, and robust fail-safe mechanisms that allow users to quickly regain control. The UX must support a collaborative relationship between human and AI, where the AI's autonomy is balanced with human judgment and control. For instance, an AI-powered robotic assistant in a manufacturing setting might autonomously optimize its movements for efficiency, but the UX should allow a human supervisor to easily override its decisions if safety concerns arise or if a deviation from planned production is necessary.

Designing for AI also requires a deeper consideration of the user's mental model of the system. With traditional software, users often develop a fairly accurate mental model of how the system works, based on its interface and consistent behavior. With AI, this mental model can become more complex and potentially inaccurate due to the system's learning and adaptive capabilities. Users might attribute intent or understanding to the AI that it does not possess, leading to frustration or misunderstanding. The UX design must therefore actively work to shape and maintain an accurate mental model. This can be achieved through clear communication about the AI's capabilities and limitations, explicit labeling of AI-generated content versus human-generated content, and providing feedback mechanisms that help users understand how their interactions influence the AI's behavior. For example, a chatbot designed to assist with customer service should clearly state that it is an AI and not a human, and its responses should be designed to reflect its current knowledge base rather than implying genuine sentience.

Furthermore, the iterative nature of AI development and deployment necessitates a continuous feedback loop in the UX design process. Unlike traditional software where major updates might be infrequent, AI systems are often continuously learning and evolving. This means that the UX must be designed to gather ongoing feedback from users about their experience with the evolving system. This feedback is not just for bug reporting; it is crucial for understanding how the AI's adaptations are perceived, identifying new usability issues that arise from its learning, and informing future development. This requires robust analytics, user surveys, A/B testing of different AI behaviors, and direct user interviews. The UX team must become adept at analyzing this continuous stream of data to refine the AI's interactions and ensure that it remains aligned with user needs and expectations.

Consider the implications for error handling. In traditional design, errors are often predictable and have well-defined recovery paths. With AI, errors can be more subtle, emergent, and harder to predict. An AI might produce a subtly biased recommendation, or misinterpret a user's intent in a way that leads to an incorrect but plausible output. The UX for error handling in AI systems must therefore be more nuanced. It should focus not just on fixing immediate errors, but on helping users understand *why* an error occurred, providing them with ways to correct the AI's understanding, and preventing similar errors in the future. This might involve designing "undo" functionalities that allow users to reverse the AI's actions, "correction" interfaces where users can provide explicit feedback about incorrect outputs, and "clarification" prompts where the AI asks for more information when it is uncertain. The goal is to empower users to co-manage the AI's learning and mitigate potential negative outcomes, fostering a sense of agency rather than helplessness.

The design of conversational AI interfaces offers a particularly rich area for exploring these shifts. Voice assistants and chatbots are prime examples of AI systems where UX is paramount. Users interact with these systems using natural language, a modality that is inherently ambiguous and context-dependent. Designing for these interfaces requires deep understanding of human conversation, pragmatics, and the nuances of natural language processing. A successful conversational UX is one that feels natural, anticipates user needs, handles ambiguity gracefully, and provides clear feedback. This often involves designing for a spectrum of AI capabilities – from simple task completion to more complex, open-ended dialogue – and ensuring that the interface clearly communicates what the AI is capable of at each stage. The UX designer must anticipate potential conversational breakdowns and design strategies for recovery, such as providing helpful suggestions for what the user can say next or offering to transfer the user to a human agent.

Moreover, the ethical considerations inherent in AI design have profound UX implications. As we've discussed, AI can perpetuate or even amplify societal biases. The UX designer has a critical role in mitigating these biases through thoughtful interface design. This can involve ensuring that AI-generated content is presented neutrally, providing users with options to report biased outputs, and designing interfaces that promote fairness and inclusivity. For example, an AI recruitment tool should not only be fair in its algorithms but also present candidate information in a way that minimizes unconscious bias on the part of the human reviewer. Similarly, a content recommendation system should be designed to avoid creating filter bubbles or exposing users to harmful or extremist content, even if the underlying AI algorithms might inadvertently do so. The UX becomes a safeguard, a point of human intervention and control in an otherwise automated process.

The concept of "delight" in UX also takes on new dimensions with AI. While traditional design strives for efficiency and satisfaction, AI can unlock entirely new forms of user delight. Imagine an AI that can generate personalized stories, compose music based on a user's mood, or create unique visual art on demand. These are experiences that go beyond mere utility; they tap into creativity, novelty, and emotional engagement. Designing for such experiences requires an understanding of human psychology, creativity, and the desire for personalized, emergent content. The UX must facilitate these creative collaborations between human and AI, making the process intuitive and rewarding, and allowing for serendipitous discovery.

Finally, the fundamental shift in designing for AI lies in embracing uncertainty and empowering users within that uncertainty. Traditional UX aimed to minimize uncertainty for the user. AI design, however, must acknowledge that a certain degree of uncertainty is inherent in intelligent systems. The goal then shifts from eliminating uncertainty to managing it effectively. This means providing users with the right information at the right time, offering clear controls, and building trust through transparency and reliability. It means designing for a partnership between humans and intelligent machines, where each leverages its strengths to achieve better outcomes. The UX practitioner, therefore, becomes less of an architect of fixed pathways and more of a facilitator of dynamic, intelligent interactions, ensuring that the human remains at the center of the experience, even as the intelligence evolves around them. This requires a continuous exploration of human-AI symbiosis, pushing the boundaries of what's possible while always grounding the design in human needs, values, and aspirations.

The advent of artificial intelligence has ushered in a new era for user experience (UX) design, shifting the focus from predictable, deterministic interfaces to dynamic, adaptive, and often opaque systems. This evolution necessitates a profound re-evaluation of how we craft human-AI interactions, moving beyond established usability principles to embrace the inherent complexities of intelligent agents. At the heart of this transformation lies the critical need to address the inherent opaqueness of many AI systems. Advanced models, particularly those based on deep learning, can function as "black boxes," making it exceedingly difficult for users—and sometimes even developers—to comprehend the precise reasoning behind a specific output, recommendation, or decision. This lack of transparency is not merely an academic concern; it is a significant barrier to user adoption, trust, and effective collaboration with AI.

Building trust is paramount when users interact with AI systems that influence critical aspects of their lives, from financial decisions and healthcare recommendations to content consumption and automated task execution. When an AI system, for instance, denies a loan application, flags a potential security threat, or suggests a particular course of action in a complex scenario, users are not only entitled to understand the outcome but also the underlying rationale. Without this understanding, users can feel powerless, suspicious, or disengaged, ultimately undermining the potential benefits of the AI. UX design, therefore, must proactively integrate mechanisms that shed light on the AI's decision-making processes. This doesn't invariably mean exposing the intricate mathematical formulas or the millions of parameters within a neural network. Instead, it entails translating complex AI operations into user-understandable explanations that are both informative and actionable.

The goal of designing for transparency and explainability is multifaceted. Firstly, it is about empowering users. When individuals understand why an AI has made a particular suggestion, they can make more informed decisions about whether to accept, reject, or modify that suggestion. This is particularly crucial in domains like healthcare, where a diagnostic AI might suggest a treatment plan. A physician, armed with an explanation of the AI's reasoning—perhaps highlighting specific symptoms, lab results, or historical patient data that strongly correlated with a particular condition—can better evaluate the AI's recommendation, discuss it with the patient, and ultimately arrive at a more robust and personalized care strategy. Similarly, in a financial context, an AI might recommend a particular investment strategy. Transparency allows the user to understand the perceived risks and rewards, the market conditions considered, and the user's own financial profile that informed the recommendation, enabling them to proceed with confidence or seek further clarification.

Secondly, transparency and explainability are vital for debugging and improving AI systems. When an AI produces an undesirable or incorrect output, understanding
why it did so is essential for developers and designers to diagnose the problem. This could stem from biased training data, flawed model architecture, or misinterpretation of input. By providing designers and developers with insights into the AI's reasoning process, it becomes easier to identify these issues and implement corrective measures. For the end-user, this translates into a more reliable and trustworthy system over time, as their feedback and observations, facilitated by transparent interfaces, can directly contribute to the AI's refinement.

Several techniques can be employed to achieve transparency and explainability in AI UX design, tailored to the specific AI application and the target audience. One common approach is to provide justifications for recommendations. For instance, a content recommendation engine—whether for movies, news articles, or products—can go beyond simply presenting a list of suggested items. It can accompany each recommendation with a brief explanation, such as "Because you watched [Movie Title X]" or "Users who liked [Article Y] also found this interesting." This not only clarifies the AI's logic but also provides users with an opportunity to refine their preferences by indicating whether the justification is accurate or not. For example, if the AI suggests a horror movie based on a user's past preference for thrillers, and the user explicitly dislikes horror, this feedback can be used to adjust the AI's understanding of their taste.

Visualizing decision processes offers another powerful avenue for transparency. For AI systems that operate through a series of steps or weigh multiple factors, graphical representations can make the complex intelligible. Consider an AI-powered resume screening tool. Instead of merely presenting a ranked list of candidates, it could visually highlight the keywords, skills, and experience that matched the job description for each applicant. This could take the form of a "match score" that breaks down into different categories, or a word cloud emphasizing the most relevant terms. For a loan application, a visual representation might show how factors like credit history, income, and debt-to-income ratio contributed to the approval or denial decision, perhaps using bar charts or radar plots to illustrate the relative importance of each factor.

The level of detail provided in these explanations is a critical design consideration. Not all users require or desire the same depth of information. A novice user interacting with a personal finance AI might be satisfied with a high-level explanation of why a particular budget adjustment is recommended, focusing on the immediate impact on their savings goals. Conversely, a financial analyst using a sophisticated market prediction AI would likely need a more granular explanation, delving into statistical models, confidence intervals, and the sensitivity of predictions to various market inputs. Therefore, a layered approach, often referred to as "progressive disclosure," is highly effective. This allows users to start with a concise, easily digestible explanation and then drill down into more detailed information if they choose. This might involve a primary explanation that summarizes the key reasons, with links or expandable sections that reveal more technical details, data sources, or alternative interpretations.

Furthermore, designing for explainability involves carefully managing user expectations about the AI's capabilities and limitations. It is crucial to avoid anthropomorphizing the AI in a way that suggests consciousness, sentience, or human-like understanding. Clear labeling and explicit communication can help in this regard. For instance, conversational agents should clearly identify themselves as AI and not pretend to be human. When an AI makes a mistake, the system should be designed to acknowledge its limitations and provide pathways for correction, rather than making excuses or attempting to mislead the user. This builds a foundation of honesty and reliability, which are cornerstones of trust.

The concept of "explainable AI" (XAI) is a burgeoning field that directly addresses these challenges. XAI research focuses on developing methodologies and tools to make AI systems more understandable. From a UX perspective, this translates into incorporating XAI techniques into the interface design. This might involve integrating LIME (Local Interpretable Model-agnostic Explanations) or SHAP (SHapley Additive exPlanations) values into the user interface, presenting them in a way that is comprehensible to the intended audience. For instance, SHAP values, which attribute the contribution of each feature to the model's prediction, could be translated into statements like "Your application was approved primarily because of your strong credit history, which contributed X points to your score. Your stable employment history also played a significant role, adding Y points."

The design of feedback mechanisms is intrinsically linked to transparency. When users can not only understand the AI's reasoning but also provide feedback on whether that reasoning is correct or helpful, they become active participants in the AI's learning and refinement process. This could involve simple "thumbs up/thumbs down" options on recommendations, or more detailed forms where users can elaborate on why they disagree with an AI's output. This iterative loop of explanation, interaction, and feedback is essential for building user confidence and ensuring that the AI's behavior remains aligned with user intent and societal values. For example, if an AI-powered image recognition system incorrectly identifies an object, providing a user interface that allows the user to correct the label and optionally explain the correction can help retrain the model and prevent future errors.

The context of the AI's application significantly dictates the nature of transparency required. In high-stakes domains like autonomous driving, transparency is not just about explaining a decision after it's made, but also about communicating the AI's current state and intentions in real-time. Dashboards that visualize sensor data, predicted trajectories, and the AI's confidence levels can provide drivers with the situational awareness needed to trust and, if necessary, override the system. Similarly, in medical applications, transparency might involve presenting not only the AI's diagnosis but also the confidence score associated with that diagnosis and any alternative diagnoses that were considered. This empowers clinicians to exercise their professional judgment more effectively.

The challenge of transparency is further complicated by

the fact that the AI's decision-making process can be dynamic and context-dependent. What might be a relevant factor for one decision might be irrelevant for another. Therefore, explainability mechanisms need to be equally adaptive. An AI that adapts its recommendations based on time of day, user location, or ongoing events needs to be able to explain its current reasoning in light of these evolving factors. For example, a smart home AI might adjust lighting based on detected natural light levels and the time of day. Its explanation for a particular lighting setting could be "Dimming the lights because it's evening and the ambient light is low."

Moreover, designing for transparency can also help in identifying and mitigating AI biases. When users understand the factors influencing an AI's output, they are better equipped to spot instances where those factors might be unfairly weighted, leading to discriminatory outcomes. For example, if an AI hiring tool consistently favors candidates from certain demographic groups, and its explanations reveal a disproportionate weighting of factors historically correlated with those groups (e.g., specific alma maters or previous employers), this provides a clear signal of potential bias that can then be investigated and addressed. The UX designer's role here is to surface these potential issues in an accessible way, prompting scrutiny and enabling corrective action.

In summary, designing for transparency and

explainability is no longer an optional feature for AI-powered systems; it is a fundamental requirement for building user trust, fostering effective interaction, and ensuring ethical development. By employing techniques such as providing clear justifications, visualizing decision processes, adopting progressive disclosure, managing expectations, and integrating feedback mechanisms, UX designers can transform opaque AI systems into understandable and empowering tools. This shift not only enhances the user experience but also lays the groundwork for more responsible and beneficial integration of AI into our lives. The ability to understand, and to be understood by, intelligent systems is the bedrock upon which a truly human-centric AI future will be built.

The introduction of artificial intelligence into user interfaces brings with it a unique set of challenges, particularly in the realm of managing user expectations. Unlike traditional software, which operates on predictable, deterministic logic, AI systems possess an inherent capacity for adaptation, learning, and sometimes, emergent behavior. This dynamism, while offering powerful capabilities, also introduces a degree of uncertainty that can be a fertile ground for misaligned expectations. Users, accustomed to the predictable responses of conventional applications, might project their understanding of human-like intelligence onto AI, leading to scenarios where the AI's actual performance falls short of what the user implicitly or explicitly anticipated. This gap between perceived and actual capability can manifest as user frustration, a decline in trust, and ultimately, a failure to adopt or effectively utilize the AI system.

A critical aspect of managing user expectations begins with clear and honest communication about the AI's intended purpose and scope from the very outset of the user's interaction. This is particularly vital during the onboarding process. For instance, consider a new smart home device powered by AI. Instead of presenting it as an all-knowing assistant capable of anticipating every need, the onboarding flow should meticulously delineate what the AI
can do and, just as importantly, what it *cannot*. For example, a smart thermostat might be advertised as learning user preferences to optimize energy consumption. The onboarding could explain that it learns by observing manual adjustments over time, that its learning rate might vary, and that it cannot predict sudden changes in occupancy or unexpected weather events without explicit input. This upfront clarification sets a realistic baseline, preventing the user from assuming a level of predictive omniscience that the system simply does not possess. Similarly, a generative AI art tool should clearly state that its outputs are interpretations based on vast datasets and prompts, and that artistic intent and nuanced emotion are concepts it emulates rather than truly possesses.

Beyond the initial introduction, ongoing feedback mechanisms play a crucial role in keeping user expectations aligned with the AI's performance. AI systems, especially those that learn and adapt, can exhibit evolving behaviors. Without clear indicators of

this evolution, users might become confused by changes in the AI's responses or recommendations. Providing real-time feedback on the AI's progress and the factors influencing its decisions can bridge this understanding gap. For a personalized news aggregator, for example, simply presenting a list of articles is insufficient. A more effective approach would be to provide subtle cues about
why an article was surfaced, such as "Recommended for you because you read about [topic X]" or "Trending in your network." This not only offers a glimpse into the AI's reasoning but also allows users to implicitly validate or correct the AI's understanding of their interests. If the AI consistently recommends articles on topics the user has no interest in, this visible discrepancy serves as an immediate signal for the user to adjust their engagement or, if the interface allows, provide direct feedback to refine the AI's learning.

Furthermore, the design of "failure states" or moments when the AI encounters limitations is paramount. Unlike traditional software that might throw an error message, AI systems can falter in more subtle, yet equally disruptive ways. A poorly designed failure response can leave a user feeling bewildered or even deceived. The goal should be to design these moments as opportunities for clarification and user intervention, rather than as dead ends. If a conversational AI cannot understand a user's request, a simple "I don't understand" is insufficient. A better approach would be to provide helpful prompts, such as "I'm still learning. Could you try rephrasing that?" or offering a menu of related options that the AI *can* process. For an AI-powered customer service chatbot, if it cannot resolve a complex issue, it should clearly and gracefully transition the user to a human agent, explicitly stating, "I've reached the limit of my capabilities on this matter. I will now connect you with a specialist who can assist you further." This acknowledges the AI's limitations while assuring the user that their needs will still be met, thereby managing expectations about the AI's capacity for problem-solving.

A significant pitfall in designing AI interactions is the temptation to anthropomorphize the system to an extent that misleads users about its true nature. While a certain degree of conversational naturalness can enhance user experience, imbuing the AI with qualities it does not possess—such as consciousness, emotions, or genuine intent—can lead to profound misunderstandings. For example, a virtual assistant that uses overly empathetic language or claims to "feel" or "care" might lead users to believe it has a level of sentience that it does not. This can create unrealistic expectations about the AI's ability to provide emotional support or exercise moral judgment. Instead, the design should focus on mimicking helpfulness and responsiveness without crossing the line into deceptive personification. Clear disclaimers, explicit identification as an AI, and a focus on functional assistance rather than simulated emotional connection are key.

Consider the design of a smart writing assistant that offers suggestions for tone, style, and grammar. While it can be incredibly useful, it's crucial that the interface does not present these suggestions as definitive pronouncements of "correct" writing. Instead, the AI should be positioned as a collaborative tool. Suggestions could be presented with probabilistic language, such as "This phrasing *might be perceived* as more formal" or "Consider this alternative for *potentially* clearer communication." The system should also offer explanations for its suggestions, linking them to established linguistic principles or stylistic conventions, rather than framing them as the AI's personal opinion. This empowers the user to retain agency and make informed decisions, understanding that the AI is a sophisticated tool to aid their process, not an infallible arbiter of language.

The concept of "over-promising and under-delivering" is a recurring theme when expectations are not managed effectively. This can occur not only through marketing but also through the internal design of the AI itself. If an AI system is designed to present a highly confident demeanor, even when its underlying confidence in its output is low, users will naturally develop a higher expectation of its accuracy. Conversely, an AI that frequently expresses uncertainty or hedges its responses excessively can lead to user disengagement, making it seem unreliable or incompetent. The challenge lies in finding a delicate balance, communicating the AI's confidence levels in a way that is informative without being paralyzing. Visual cues, such as confidence scores displayed alongside recommendations, or qualitative labels like "highly likely," "possible," or "uncertain," can help users gauge the AI's certainty.

For instance, in a medical diagnostic AI, presenting a diagnosis with a 99% confidence score might lead a physician to rely on it too heavily, potentially overlooking subtle cues that contradict the AI's conclusion. A more responsible design might present the primary diagnosis alongside a confidence score, but also flag any differential diagnoses that the AI considered and their associated probabilities, along with a clear indication of the AI's limitations in interpreting nuanced clinical presentations or subjective patient reports. This fosters a partnership between the AI and the human expert, where the AI provides data-driven insights but does not usurp the human's critical judgment.

The design of the AI's learning process itself also impacts user expectations. If a system claims to "learn" but its behavior remains static or its improvements are imperceptible, users will quickly become disillusioned. Transparency about the learning process is essential. This could involve providing users with insights into what kind of data is being used to train the AI, how often it is updated, and what the expected outcomes of these updates are. For example, a personalized music streaming service could offer a dashboard that shows the user their "listening profile," detailing the genres, artists, and moods that the AI has identified. It could also provide an option to "retrain" the AI based on a broader selection of liked or disliked songs, making the learning process more tangible and controllable for the user. This active involvement in the AI's development can foster a sense of ownership and patience, as users understand that their interaction contributes to the system's improvement.

Furthermore, the context in which the AI operates significantly influences how expectations should be managed. An AI used for entertainment, like a game companion, can afford to be more playful and less rigidly accurate than an AI deployed in a financial advisory role. In the latter case, absolute clarity, accuracy, and the explicit communication of risks are paramount. The onboarding for a robo-advisor, for instance, must be exceptionally detailed, outlining not only the investment strategies employed but also the inherent risks associated with market fluctuations, the fees involved, and the fact that the AI cannot guarantee returns or protect against losses. Legal disclaimers are important, but they should be integrated into the user experience in a digestible manner, perhaps through interactive elements that quiz the user on their understanding of key concepts before proceeding.

The iterative nature of AI development means that even the most carefully managed expectations can be challenged over time. As AI capabilities advance, users' baseline expectations will naturally shift. UX designers must therefore adopt a philosophy of continuous evaluation and adaptation. This involves not only gathering user feedback on the AI's performance but also staying abreast of advancements in AI research and anticipating how these might influence user perceptions and desires. A proactive approach might involve periodically updating onboarding materials or providing in-app notifications about new features or enhancements to the AI's capabilities, always framed within realistic parameters.

Ultimately, managing user expectations with AI is an ongoing dialogue. It requires a commitment to honesty, transparency, and a deep understanding of human psychology. By carefully crafting initial introductions, providing continuous feedback, designing graceful failure modes, avoiding deceptive anthropomorphism, and clearly communicating limitations, UX designers can foster trust and ensure that users can harness the full potential of AI without succumbing to disappointment or misunderstanding. This thoughtful approach transforms AI from a potentially bewildering technology into a reliable and valuable partner in the user's experience.

When an artificial intelligence system encounters a situation it cannot process, misunderstands a user's intent, or generates an output that is factually incorrect or undesirable, the ensuing "error" is often more complex than a simple software bug. These are not typically crashes or freezes, but rather failures in comprehension, reasoning, or generation. The UX design for these AI-driven interactions must therefore be meticulously crafted to anticipate, communicate, and recover from these nuanced failures, ensuring that the user's journey remains as smooth and productive as possible. The goal is not merely to inform the user that something has gone awry, but to actively guide them towards resolution and reinforce their confidence in the system's overall utility.

One of the foundational principles in designing for AI errors is clear and informative communication. When an AI fails, a generic "An error occurred" message is not only unhelpful but can actively erode user trust. Instead, the system should strive to provide contextually relevant feedback. For instance, if a conversational AI is unable to parse a user's request, it should explain *why* it struggled, if possible, and suggest actionable alternatives. Consider a travel planning AI that, when asked to book a flight for a specific date, responds, "I'm sorry, I couldn't find any flights for that exact date. Would you like me to search for flights on nearby dates, or perhaps a different destination?" This type of message does two crucial things: it explains the limitation (inability to find flights on the specified date) and offers concrete next steps that the user can readily take. This is far more empowering than a cryptic error. Similarly, if a generative AI produces an image that is visually distorted or deviates significantly from the prompt, it should ideally offer a rationale, such as, "The prompt was ambiguous regarding the style of the background," and suggest ways to refine the prompt, like "Try specifying 'photorealistic' or 'painterly' for the background." This transparency demystifies the AI's internal processes and guides the user towards a successful outcome.

Beyond simple explanations, providing users with intuitive mechanisms to correct or guide the AI is paramount for robust error recovery. Many AI systems, especially those involving generative tasks or complex decision-making, will inevitably produce outputs that are not perfectly aligned with user intent. In these scenarios, the user should not be forced to start the entire process over. Instead, interfaces should offer direct manipulation or feedback loops that allow for iterative refinement. For example, if a language translation AI provides a slightly awkward phrasing, a well-designed interface would allow the user to highlight that specific phrase and offer alternative suggestions, or even directly edit it. This "in-line editing" of AI output is incredibly powerful. Likewise, in a recommendation system where a user is presented with a list of suggested products or content, the ability to explicitly downvote or provide a reason for dissatisfaction ("Not interested in this genre," "Already own this item") allows the user to directly correct the AI's understanding, leading to more accurate future recommendations. This is a form of active learning facilitated by the user, transforming a perceived error into a learning opportunity for the AI and a point of control for the user.

The concept of "recovery paths" in AI interactions is critical to minimizing disruption and rebuilding user confidence after an error. When an AI falters, the recovery process should aim to restore the user to a state where they can continue their task with minimal frustration and a renewed sense of trust. This might involve rolling back to a previous, stable state, offering a simplified alternative mode of operation, or gracefully escalating to a human or a more robust system. For instance, if an AI-powered assistant is engaged in a multi-step task, and it encounters an error midway, it should be able to indicate precisely where it left off and provide options to either retry the problematic step, skip it entirely if possible, or revert to the last successfully completed step. Imagine a complex data analysis AI that fails during an intermediate calculation. A good recovery would involve reporting the specific calculation that failed, offering to re-run that calculation with adjusted parameters, or suggesting that the user manually input the result for that step. This proactive approach to recovery prevents the user from feeling lost or that their previous efforts have been in vain.

Furthermore, the design of error states can also serve as an opportunity to educate users about the AI's capabilities and limitations, thereby managing expectations and preventing future misunderstandings. When an AI explicitly states what it *cannot* do in response to a request, it’s not just an error message; it’s a boundary definition. For example, if a financial AI is asked for investment advice on highly speculative assets, and it responds with, "I am designed to provide guidance on diversified, long-term investment strategies. I cannot offer advice on highly volatile or speculative assets due to regulatory and risk management protocols," this informs the user about the AI's scope and ethical guardrails. This proactive disclosure of limitations, even in the context of an inability to fulfill a request, is a form of error handling that empowers users to interact more effectively with the AI in the future. It transforms potential frustration into a learning experience.

The nature of AI errors can also be subtle, particularly in systems that learn and adapt over time. Sometimes, the "error" is not a single, distinct event, but a gradual drift in performance or a deviation from expected behavior that a user might not immediately recognize as an error. For example, a personalized content feed might begin to show less relevant articles due to shifts in user behavior or changes in the underlying algorithms that are not transparent to the user. In such cases, error handling transitions from reactive messaging to proactive monitoring and user-driven recalibration. Designing systems that allow users to periodically review and "reset" their preferences, or that provide summaries of *why* certain content is being recommended, can help users identify and correct these subtle performance degradations. Offering a feature like "Show me why this is recommended" or "Adjust my preferences for this category" allows users to self-correct the AI's trajectory, effectively handling emergent "errors" of irrelevance.

In instances where an AI generates incorrect factual information, the recovery strategy needs to be particularly robust, emphasizing accuracy and user agency. This is especially critical in domains like education, healthcare, or research. When an AI makes a factual error, it should be able to acknowledge the mistake, provide the correct information, and ideally, explain how the error occurred to prevent recurrence. For example, a historical fact-checking AI that provides an incorrect date for an event should ideally offer a correction upon user query or feedback, accompanied by a link to its source or an explanation of the misinterpretation. The interface should make it easy for users to report inaccuracies, perhaps through a simple "Is this correct?" button or a feedback form. The AI's response to such reports is crucial: a simple acknowledgment is good, but offering a corrected answer and a brief explanation of the fix demonstrates a commitment to accuracy and recovery.

The design of AI error handling must also consider the emotional state of the user. Errors can be frustrating, and a poorly handled error can lead to anger and abandonment of the system. Therefore, the tone and language used in error messages are as important as the content. Empathetic language, even from a non-sentient AI, can go a long way. Instead of a cold, technical message, an AI might say, "I apologize, I seem to have misunderstood your request. Let's try that again," or "I'm having trouble processing that. Could you please rephrase it for me?" This human-like communication style can de-escalate frustration and encourage continued interaction. The visual design of error states also plays a role; clear, uncluttered error screens with prominent calls to action for recovery are more effective than confusing or overwhelming displays.

When an AI system is deeply integrated into a workflow, a complete system failure or a critical error can have significant repercussions. In such high-stakes scenarios, the recovery mechanisms must be exceptionally resilient. This might involve having fallback systems in place, maintaining offline capabilities, or providing clear protocols for manual intervention. For instance, an AI system used in critical infrastructure management might have redundant AI models or a direct human override feature that can be activated instantly if the primary AI experiences an unrecoverable error. The user interface for such a recovery scenario needs to be intuitive and accessible, even under pressure, ensuring that operators can quickly take control and mitigate any potential risks. The process of switching to a manual or backup system should be clearly signaled, providing users with confidence that the operation remains under control.

Ultimately, effective error handling and recovery in AI interactions are not about eliminating all errors—an unrealistic goal, especially with complex AI—but about building systems that are forgiving, transparent, and user-centric. By embracing a design philosophy that anticipates failure, communicates it clearly, provides intuitive correction mechanisms, and offers robust recovery paths, UX designers can ensure that even when the AI stumbles, the user's experience remains positive and productive. This approach fosters a relationship of trust and reliability, enabling users to confidently engage with AI technologies, knowing that the system is designed to support them through challenges, not just when everything goes perfectly. The ability of an AI to gracefully recover from an error often becomes a defining characteristic of its perceived intelligence and usability, transforming potential moments of frustration into opportunities for enhanced user engagement and learning.

The advent of artificial intelligence has ushered in a new era of user experience, moving beyond static, one-size-fits-all interfaces to dynamic, deeply personalized, and adaptive interactions. At the heart of this transformation lies the AI's capacity to learn, understand, and respond to individual users in ways previously unimaginable. This evolution is not merely about convenience; it's about creating experiences that feel intuitive, relevant, and ultimately, more valuable to each person interacting with the technology.

One of the most profound ways AI enhances UX is through personalization. Unlike traditional software that presents a uniform interface to all users, AI-driven systems can dynamically tailor their presentation, functionality, and content to match the unique preferences, behaviors, and needs of each individual. This begins with the AI's ability to infer and learn user preferences. Through observing patterns of interaction—what a user clicks on, how long they spend on certain content, what queries they make, their past choices, and even implicit feedback like scrolling speed or dwell time—AI algorithms can build a sophisticated profile of the user. This profile isn't static; it's a living document that evolves with every interaction. For example, a news aggregation app might notice that a user consistently spends more time reading articles about technological advancements and rarely engages with sports news. The AI would then prioritize technology-related stories in the user's feed, perhaps even suggesting deeper dives into specific sub-topics within technology that the user has shown interest in. This ensures that the user's limited attention is directed towards content they are most likely to find valuable, fostering engagement and reducing the cognitive load of sifting through irrelevant information.

This personalization extends beyond content delivery to the very interface and functionality of the application. Consider an AI-powered productivity suite. An AI could learn that a particular user frequently uses a specific set of advanced formatting tools in a document editor. Instead of requiring the user to navigate through multiple menus to access these tools, the AI could proactively surface them, perhaps by adding them to a custom quick-access toolbar or even suggesting them as the user begins typing. Similarly, in a customer relationship management (CRM) system, an AI might learn that a sales representative tends to follow a particular sequence of actions when contacting a new lead. The AI could then guide the user through this workflow, presenting the relevant information and suggesting the next logical steps, thereby streamlining their tasks and reducing the chance of forgetting critical follow-up actions. This level of personalization makes the technology feel less like a generic tool and more like a bespoke assistant, perfectly attuned to the user's working style.

Adaptability is the natural counterpart to personalization. If personalization is about tailoring the experience to the individual
at a given moment, adaptability is about the system's ability to change and evolve *over time* as the user's needs, skills, or preferences shift. This dynamic evolution is crucial for maintaining relevance and usability throughout a user's relationship with a system. Adaptive learning platforms are a prime example of this principle in action. These educational systems use AI to continuously assess a student's understanding and progress. If a student struggles with a particular concept, the AI can automatically provide supplementary materials, offer simpler explanations, or present practice problems at a lower difficulty level. Conversely, if a student demonstrates mastery, the AI can accelerate their progress, introduce more challenging material, or provide opportunities for deeper exploration. This creates a learning journey that is uniquely paced and contoured to each student's learning curve, maximizing retention and minimizing frustration or boredom.

Beyond education, adaptability is vital in many AI-powered applications. Imagine a virtual assistant that initially learns a user's basic commands and common requests. As the user becomes more proficient and their needs grow more complex, the AI should be able to adapt its understanding and offer more sophisticated functionalities. For instance, after a user has repeatedly asked the assistant to play a specific playlist at a certain time of day, the AI might proactively suggest setting a recurring alarm with that playlist. Or, if the assistant notices a pattern of the user asking for weather forecasts before a specific type of outing (e.g., hiking), it could begin to offer weather alerts tailored to those excursions without explicit prompting. This proactive adaptation demonstrates an AI that is not just responding, but anticipating and evolving alongside the user.

The design considerations for personalization and adaptability are multifaceted. Firstly, designers must determine
what aspects of the experience are suitable for personalization. Not every feature benefits from being individually tailored; sometimes, a consistent experience across all users is desirable for clarity or collaboration. Once the scope is defined, the next challenge is *how* the AI will learn these preferences. This involves selecting appropriate machine learning models and data sources, ensuring that the learning process is efficient and accurate. Crucially, users need to have visibility and control over this personalization. A "black box" AI that changes behavior without explanation can be disorienting. Therefore, interfaces should ideally offer mechanisms for users to review their inferred preferences, correct misunderstandings, and explicitly set or override personalized settings. For example, a streaming service might allow users to see a list of genres they are "interested in" and provide

buttons to add or remove them, or even to indicate why they might be interested (e.g., "suggested by a friend").

Furthermore, designing for adaptability requires foresight. The system must be architected to accommodate evolving user behaviors and AI model updates. This often involves a modular design where different components of the AI can be updated or retrained independently. A key aspect of adaptability is graceful degradation; if the AI's adaptive capabilities are temporarily limited or its learning falters, the system should still function reliably, perhaps reverting to a more basic or default mode, rather than failing entirely. This ensures a robust and dependable user experience, even as the AI continuously learns and adapts.

The ethical implications of personalization and adaptability are significant and demand careful consideration. The very mechanisms that enable tailored experiences rely on collecting and processing user data, raising concerns about privacy. As AI systems gather more intimate details about users—their habits, interests, and even emotional states inferred from interaction patterns—robust data protection measures become paramount. Users must be informed about what data is being collected, how it is being used to personalize their experience, and have clear control over their data, including the ability to opt out of certain types of data collection or personalization. Transparency regarding data usage is not just an ethical imperative but also a legal requirement in many jurisdictions.

Another critical ethical challenge is the potential for "filter bubbles" or "echo chambers." When AI aggressively personalizes content, it risks showing users only information and perspectives that align with their existing beliefs and preferences. This can lead to a narrowing of worldview, making individuals less exposed to diverse ideas and potentially reinforcing biases. For instance, a news aggregator that exclusively shows articles confirming a user's political leanings, while omitting counterarguments or alternative viewpoints, can contribute to societal polarization. Designers must actively work to mitigate this by incorporating mechanisms that introduce serendipity, expose users to diverse perspectives, or provide clear labeling of personalized content. This could involve offering a "discovery" mode that intentionally surfaces less familiar topics or viewpoints, or clearly indicating when content has been prioritized based on learned preferences, allowing users to consciously seek out broader information.

The concept of "fairness" is also intertwined with personalization. If an AI system's personalization algorithms inadvertently discriminate against certain user groups, it can lead to inequitable outcomes. For example, an e-commerce recommendation engine that personalizes product suggestions might, due to biased training data, consistently recommend lower-priced items to users from certain demographic backgrounds, limiting their access to potentially desirable products. Ensuring that personalization algorithms are audited for bias and that systems are designed to promote equitable access to information and opportunities is crucial. This involves not only technical solutions but also a conscious effort to consider the societal impact of AI-driven personalization.

Furthermore, the opacity of some AI learning processes can make it difficult for users to understand *why* certain recommendations are made or why the system behaves in a particular way. This lack of explainability can erode trust, especially when the personalized output is unexpected or perceived as incorrect. Designing for explainable AI (XAI) in personalized systems is therefore important. Providing users with simple, understandable explanations for personalized recommendations—such as "You're seeing this because you recently viewed similar items" or "This article is trending among users with similar interests"—can foster transparency and empower users to better interpret and interact with the personalized experience.

The ultimate goal of personalization and adaptability in AI UX is to create systems that are not just functional, but profoundly human-centered. By leveraging AI's learning capabilities, designers can craft experiences that feel intuitive, supportive, and tailored to the individual's journey. This requires a delicate balance: harnessing the power of AI to understand and adapt while remaining mindful of ethical responsibilities, ensuring privacy, fostering diverse perspectives, and maintaining user agency and trust. When executed thoughtfully, personalization and adaptability transform technology from a mere tool into a truly collaborative partner, enhancing productivity, learning, and engagement in ways that are deeply meaningful to each user. The ongoing evolution of AI promises even more sophisticated forms of personalized and adaptive experiences, making it an ever more critical area for UX design and ethical consideration. The landscape of user interfaces is shifting, moving from static blueprints to living, breathing digital companions that learn and grow with us, making our interactions with technology more efficient, enjoyable, and ultimately, more human.

Consider the evolution of a smart home ecosystem. Initially, a user might set up basic controls for lights and temperature. An AI-powered system would then begin to learn patterns: when the user typically wakes up, when they leave for work, their preferred ambient temperature for different times of day, and which lights are used most frequently. Based on this, the AI can automate these functions. It might learn to gradually brighten bedroom lights in the morning, adjust the thermostat to a lower setting when the user typically leaves, and turn off all lights in the evening when the user is usually asleep. This is personalization in action, optimizing the home environment to the user's daily rhythm.

The adaptability aspect comes into play as the user's life changes. Perhaps the user starts working from home. The AI, noticing a shift in their departure patterns and increased activity during traditional work hours, would adapt its automations. It might learn to maintain a comfortable temperature throughout the day, disable the "away" mode automations, and perhaps even learn to adjust lighting based on the natural daylight available, further optimizing the environment for prolonged presence. If the user begins entertaining guests more frequently, the AI could learn to adjust lighting and music settings for social gatherings, or offer quick presets for "hosting mode." This continuous learning and adjustment ensure that the smart home remains a comfortable and convenient space, evolving with the user's lifestyle without requiring constant manual reprogramming.

The ethical considerations here are clear. The AI is collecting data on the user's presence, daily routines, and preferences within their private space. This data is sensitive. Therefore, strong encryption, secure data storage, and transparent consent mechanisms are vital. Users must be able to access the data the AI has collected about them, understand how it influences its actions, and have the ability to delete this data or reset the AI's learning. Moreover, the system should avoid creating "comfort traps." If the AI becomes so efficient at anticipating the user's every comfort and preference that it removes any need for effort or decision-making, it could inadvertently reduce user agency and lead to a passive existence. Designers must ensure that while the AI provides convenience, it still allows for intentional overrides and encourages mindful engagement with the environment, rather than pure automated existence. Introducing elements of playful discovery, like suggesting an unusual lighting scheme for an evening or proposing a new playlist based on detected moods (inferred from interaction speed or music choices), can add a touch of delightful serendipity without compromising core functionality.

In the realm of online retail, personalization and adaptability are even more pronounced. Beyond simple product recommendations, AI can personalize the entire shopping journey. This might include tailoring the layout of the website, highlighting promotions most relevant to the user's past purchasing behavior, and even adapting the language and tone of product descriptions. For instance, a user who frequently buys organic and sustainable products might see banners emphasizing eco-friendly materials and certifications, while a user focused on budget might see prominent discounts and value bundles.

The adaptive element comes into play as user needs and market trends change. If a user's purchasing habits shift from casual wear to professional attire, an adaptive system would gradually recalibrate its recommendations. If a brand the user frequently buys from releases a new product line, the AI could proactively alert the user based on their affinity for that brand. Furthermore, AI can adapt the search functionality itself. If a user consistently refines their searches with specific terms, the AI can learn to prioritize those terms in future searches, making the search experience more efficient over time.

However, the ethical tightrope walk is evident. The risk of creating exclusive shopping experiences that inadvertently limit a user's exposure to new brands or product categories is real. If an AI only shows a user products that are very similar to what they've bought before, it can stifle discovery and innovation for both the consumer and the retailer. This is where intentional design for serendipity becomes crucial. A well-designed system might periodically inject "discovery items" into a user's feed—products that are tangentially related but not direct matches, perhaps from emerging brands or different categories—with a clear label indicating that this is a "discovery suggestion." This approach balances personalization with the exploration of new possibilities. The potential for price discrimination, where users are shown different prices for the same product based on their perceived willingness to pay (inferred from their browsing and purchasing history), is another significant ethical concern that must be actively guarded against. Transparency about pricing and offering consistent pricing policies for all users is essential to maintain trust.

In the context of professional tools, personalization and adaptability can dramatically boost productivity.

Consider a software development environment. An AI could learn a developer's preferred coding languages, frequently used libraries, and common debugging patterns. It could then adapt the IDE's interface, pre-load relevant documentation, offer context-aware code suggestions, and even flag potential errors based on the developer's historical mistakes. If a developer consistently struggles with a particular type of bug, the AI could proactively offer tips or resources related to that issue when the developer starts a new project.

Adaptability ensures that as the developer masters new techniques or starts working on different types of projects, the AI's support evolves. A junior developer who gradually becomes a senior engineer will experience a corresponding evolution in the AI's assistance, moving from basic syntax help to more complex architectural suggestions. The system could also adapt to project-specific needs. If a project requires working with a less familiar database technology, the AI could offer enhanced support for that specific technology, highlighting relevant tools and documentation.

Ethically, the concern here revolves around over-reliance and skill atrophy. If an AI becomes too helpful, developers might not develop their core problem-solving skills or their deep understanding of underlying technologies. Therefore, the design should empower, not replace, the human professional. This means providing clear explanations for AI suggestions, allowing easy overrides, and framing AI assistance as a guide rather than an oracle. Transparency about the AI's learning process and its limitations is also key. Developers should understand that the AI is a tool learning from observed patterns, not an infallible expert. Auditing the AI's suggestions for correctness and ensuring that it doesn't perpetuate inefficient or incorrect coding practices learned from past interactions is also a crucial ethical safeguard. The goal is to augment human expertise, not to render it obsolete or to create a generation of developers who can only code with AI assistance.

Across all these domains, the overarching principle is to design AI systems that are not just intelligent, but also empathetic, transparent, and empowering. Personalization and adaptability, when wielded responsibly, can create digital experiences that feel more natural, more efficient, and more valuable than ever before. However, this power comes with a profound responsibility to safeguard user privacy, prevent the formation of restrictive echo chambers, ensure fairness, and maintain transparency. The future of AI UX lies in this delicate yet essential balance, where technology enhances our lives by understanding and adapting to us, while respecting our autonomy and our place in a broader, diverse world.

Chapter 7: Building Accurate Mental Models for Users

The human mind is a remarkable engine for pattern recognition and sense-making. When faced with any new system, whether it's a physical device, a complex process, or a digital interface, we don't just interact with it passively. Instead, we actively construct an internal representation of how it works, what its components are, how they relate to each other, and what the likely outcomes of our actions will be. This internal representation is known as a **mental model**. It's a cognitive framework, a simplified map of reality that allows us to navigate and predict the behavior of the world around us, including the technological tools we use.

Think about the last time you encountered a new kitchen appliance, say, a sophisticated coffee machine with multiple buttons and settings. Before even reading the manual, you likely brought to bear your existing mental models of similar devices. You'd expect certain buttons to control power, others to dispense water or coffee, and perhaps some to adjust temperature or strength. You'd anticipate that pressing a "brew" button would initiate the coffee-making process, and that the machine would require water and coffee grounds. These expectations, informed by past experiences and general knowledge about how machines operate, form your initial mental model of that specific coffee machine. The accuracy and completeness of this mental model directly influence how quickly and effectively you can use the machine. If your mental model is accurate – you understand that the large dial controls grind size and the illuminated button starts the brew cycle – you'll likely have a smooth, intuitive experience. Conversely, if your mental model is flawed – perhaps you believe the same button that dispenses hot water also makes the coffee – you'll encounter frustration, confusion, and errors.

This fundamental psychological concept of mental models is paramount when designing any interactive system, and it becomes even more critical in the context of artificial intelligence. AI systems, by their very nature, can be incredibly complex, abstract, and, to the user, often opaque. They operate based on intricate algorithms, vast datasets, and often, processes that are not directly observable. The challenge for designers is to bridge this gap between the internal workings of the AI and the user's understanding. Our goal is to facilitate the formation of an accurate mental model in the user's mind, one that aligns as closely as possible with the AI's actual functionality and capabilities.

A well-formed mental model acts as a cognitive shortcut, enabling users to interact with a system efficiently and effectively. When a user possesses an accurate mental model of an AI system, they can:

Predict System Behavior: They can anticipate what will happen when they perform a specific action. For instance, a user with a good mental model of a smart home AI might know that saying "Lights on" will

activate the pre-programmed evening lighting scene, rather than just a single lamp.
Diagnose Errors: When something goes wrong, they can often pinpoint the cause based on their understanding of the system. If their smart assistant doesn't respond, they might consider if they spoke too softly, if the microphone is muted, or if the system is offline – possibilities derived from their mental model.
Formulate Goals: They can set realistic expectations and devise strategies to achieve their objectives. Knowing the AI's capabilities allows them to decide whether to ask for a complex data analysis or a simple factual retrieval.
Learn New Features: An accurate mental model provides a foundation for understanding new functionalities. If a user understands the core principles of how an AI personalizes recommendations, they can more readily grasp why a new type of recommendation is being offered.
Feel in Control: Ultimately, having a clear understanding of how a system operates fosters a sense of agency and reduces anxiety. Users feel empowered rather than bewildered.

The absence of an accurate mental model, on the other hand, leads to a cascade of negative consequences. Users may repeatedly make the same mistakes, become frustrated by unpredictable behavior, abandon the system altogether, or develop misconceptions that are difficult to unlearn. They might attribute magical or sentient qualities to the AI, leading to unrealistic expectations and potential disappointment. This phenomenon is often observed when users interact with AI that exhibits sophisticated natural language processing or seemingly intelligent decision-making. Without a clear understanding of the underlying probabilistic models or data-driven inferences, users might anthropomorphize the AI, attributing consciousness or intentionality where only complex computation exists.

Consider the early days of voice assistants. Many users struggled with the implicit commands and context sensitivity. They might have expected the assistant to understand conversational nuances or recall past interactions spontaneously, simply because it *sounded* intelligent. When the assistant failed to meet these unarticulated expectations, users were often confused. Their mental model was likely based on science fiction portrayals of AI, rather than the reality of pattern matching, keyword recognition, and limited conversational memory. The design challenge, then, is to guide users toward an accurate mental model that reflects the *actual* capabilities and limitations of the AI.

This means designers must carefully consider what information needs to be conveyed to the user, and *how* it should be conveyed. It's not about overwhelming the user with technical details of machine learning algorithms, but rather about providing clear, consistent, and affordant cues that allow them to build a functional understanding. This often involves:

Visible Affordances: Designing interface elements that clearly indicate what actions are possible and what the expected result will be. For example, a button labeled "Analyze Data" with an icon of a chart suggests an action that will produce a visual representation of data.
Feedback Mechanisms: Providing immediate and understandable feedback after an action is taken. If a user asks an AI to perform a complex task, the system should indicate that it has received the request, is processing it, and when it expects to be done. This could be a progress bar, a spinning icon, or a textual update.
Consistent Language and Metaphors: Using familiar terminology and analogies helps users map their existing knowledge onto the new system. If an AI system categorizes information, using terms like "folders," "tags," or "categories" that users are already familiar with can greatly aid understanding.
Progressive Disclosure: Revealing complexity gradually. Users should be able to perform basic tasks easily, and only when they need more advanced functionality should they be presented with more intricate options or explanations. This prevents cognitive overload and allows users to build their mental model step-by-step.
Explanations and Transparency (where appropriate): While not always necessary for every interaction, providing accessible explanations for why an AI made a certain decision or recommendation can be invaluable. This is particularly important in domains where trust and understanding are critical, such as healthcare or finance. For instance, an AI suggesting a financial investment might explain its rationale by citing market trends or specific portfolio performance

data, allowing the user to build a more robust mental model of the AI's decision-making process.

The process of forming a mental model is iterative and dynamic. Users don't form a perfect model instantly. They interact, observe, infer, and refine their understanding over time. This is where the adaptive nature of AI can actually be a double-edged sword from a mental modeling perspective. While AI can personalize experiences to an unprecedented degree, if its adaptations are too subtle or opaque, they can prevent users from developing a stable, generalized mental model. If the interface or behavior changes constantly without clear signaling, the user's internal representation of the system becomes unstable, leading to perpetual confusion and a feeling of being out of control.

Conversely, a well-designed AI system can actively support and enhance the user's mental model. By observing how users interact, the AI can identify areas of confusion or misunderstanding. It can then proactively offer clarifying information, provide contextual help, or subtly adjust its own behavior to better align with the user's implicit model. For example, if a user repeatedly struggles to find a specific feature within an AI-powered application, the system might learn to highlight that feature more prominently for that user or offer a brief tutorial on its location and use. This isn't just personalization; it's about actively helping the user build a more accurate and useful mental model.

The concept of "affordance" is particularly relevant here, coined by psychologist James J. Gibson and popularized in design by Don Norman. Affordances are the perceived and actual properties of an object, primarily those fundamental properties which determine just how the thing could possibly be used. In user interface design, an affordance is a property that suggests how it can be used. For an AI system, this means designing interfaces and interactions that clearly communicate what the AI *can do* and *how* to elicit those actions. If an AI is capable of complex data analysis, the interface should provide clear pathways to initiate such analyses, perhaps through well-labeled buttons, intuitive command structures, or guided workflows. If the AI is designed to learn preferences, there should be explicit mechanisms for users to see what preferences have been learned and to modify them, making the learning process tangible rather than a hidden background operation.

The "knowing-doing gap" is a related psychological phenomenon that underscores the importance of mental models. It refers to the discrepancy between what people know and what they actually do. In the context of technology, a user might *know* that a certain feature exists and how it's supposed to work (they have a correct piece of knowledge in their mental model), but due to an inaccurate or incomplete overall mental model of the system's workflow, they might fail to use it effectively or even at all. Designing for accurate mental models aims to bridge this gap by ensuring that the user's understanding of the system's capabilities and interconnections aligns with the practical steps required to achieve their goals.

Ultimately, fostering accurate mental models in users is not merely a matter of usability; it's a foundational element of trust and effective human-AI collaboration. When users understand how an AI works, they are more likely to trust its outputs, collaborate with it effectively, and derive genuine value from its capabilities. Without this understanding, AI can remain a mysterious, unpredictable, and potentially intimidating force, hindering its adoption and its potential to augment human capabilities. Therefore, every design decision, from the initial conceptualization of an AI feature to the fine-tuning of its user interface, must be viewed through the lens of how it shapes the user's mental model. The goal is to make the AI feel less like an alien entity and more like a comprehensible, predictable, and ultimately, a helpful partner.

When crafting interactions with artificial intelligence, the objective is to ensure users develop a robust and accurate mental model of the system's capabilities and limitations. This is achieved by designing interfaces and dialogue flows that leverage familiar concepts and predictable patterns, thereby minimizing the cognitive load required for users to understand and operate the AI. The guiding principle is to make the AI's operations as transparent and intuitive as possible, bridging the gap between the complex underlying algorithms and the user's everyday understanding. This involves translating the AI's internal logic into a format that resonates with pre-existing user knowledge and expectations.

One of the most powerful tools in this endeavor is the use of familiar metaphors. Humans are adept at understanding new concepts by relating them to things they already know. For an AI system, this can manifest in various ways. For example, a document summarization AI could be presented as a "digital editor" that "proofreads" and "condenses" lengthy texts. This metaphor immediately conjures images of someone carefully reading, identifying key points, and then rephrasing them concisely. Similarly, a recommendation engine could be framed as a "personal shopper" or a "curator" who understands your tastes and suggests items you might like. These established roles and actions provide a readily accessible framework for users to begin building their mental model. The effectiveness of these metaphors hinges on their congruence with the AI's actual function. A "personal shopper" that consistently recommends items far outside the user's stated preferences will quickly erode trust and confuse the user's mental model, creating a mismatch between the metaphor and reality. Therefore, careful selection and validation of these metaphorical anchors are critical.

Beyond metaphors, consistency in interaction patterns is paramount. Users learn by doing, and when an AI consistently responds to similar inputs or prompts in predictable ways, users can confidently update and reinforce their mental models. This means establishing clear conventions for how users can interact with the AI and ensuring the AI adheres to these conventions. For instance, if a user is accustomed to using specific keywords to trigger certain actions in one part of an AI application, they should expect similar keywords to function analogously in other relevant areas. A deviation from this pattern, without clear explanation, can lead to confusion. Imagine an AI assistant designed for scheduling meetings. If the user can initiate a meeting request by saying "Schedule a meeting with John," they would logically expect to be able to modify an existing meeting using a similar phrasing, such as "Reschedule my meeting with John." If, instead, the AI requires a completely different, non-intuitive command for rescheduling, the user's mental model is disrupted, leading to frustration and a potential breakdown in interaction. This consistency extends to visual cues as well. If a button or an icon consistently signifies an action that leads to a particular outcome, it should continue to do so across the entire interface.

The concept of "affordances," as popularized by Don Norman, is equally vital for designing intuitive AI understanding. Affordances are the perceived and actual properties of an object that suggest how it can be used. In the context of AI, affordances are the design elements that communicate the AI's capabilities and the methods by which users can access them. For instance, a clearly labeled button that reads "Analyze Sentiment" on an AI interface immediately signals that the system can perform sentiment analysis and provides a direct means to initiate it. Similarly, an input field with a placeholder text like "Ask me anything..." or "Describe your request..." visually communicates that the system is ready to receive natural language input. When an AI is designed to learn user preferences, affordances should make this learning process transparent. This could involve providing a dedicated "Preferences" section where users can view what the AI has learned about them, and crucially, where they can edit or override these learned preferences. This explicit control mechanism acts as a powerful affordance, clarifying the AI's learning mechanism and empowering the user, thereby reinforcing an accurate mental model of the AI's adaptive nature.

Consider the design of a conversational AI, such as a customer service chatbot. To facilitate accurate mental models, the interface should provide clear visual cues about the AI's operational state. When the chatbot is actively processing a user's query, a visual indicator, such as a typing animation or a "thinking" icon, should be present. This "affords" the understanding that the AI is engaged in a task and prevents the user from assuming the system is unresponsive or has encountered an error. Conversely, when the AI is ready to receive input, the cursor should be active in the input field, and the prompt should be clear, signaling availability. If the AI has limitations in its conversational memory, this should be subtly communicated, perhaps through prompts that occasionally ask for clarification on context that a human might be expected to recall. For example, if a user has been discussing a particular product, and then asks a follow-up question about it, the AI might respond with "Regarding the Model X refrigerator you were asking about, would you like to know its energy efficiency?" This explicit re-statement of context, while perhaps slightly verbose, helps manage user expectations about the AI's memory capabilities and thus supports a more accurate mental model.

Furthermore, the design of AI systems should aim to make the underlying processes as predictable as possible, even when those processes are complex. This doesn't necessitate revealing the intricate details of machine learning algorithms, but rather providing meaningful feedback and explanations that help users infer how the AI is likely to behave. For example, an AI system that generates creative text might provide options for "creativity level" or "formality," allowing users to influence the output. The impact of these choices should be somewhat predictable; a higher "creativity level" might be associated with more imaginative or unconventional phrasing, while a lower level might result in more straightforward and conventional text. When the AI produces an output, providing brief, contextual explanations for *why* it generated that particular response can be immensely beneficial. For instance, if an AI-powered content generation tool produces a paragraph that seems slightly off-topic, it could offer a brief annotation like "This sentence was generated to connect the previous discussion on sustainable energy with the upcoming topic of renewable resource management." This "explainability" feature, even at a high level, allows users to understand the AI's reasoning and adjust their future prompts accordingly, thus refining their mental model of how the AI "thinks" and generates content.

The principle of progressive disclosure is also invaluable. Complex AI systems often possess a wide range of functionalities. Presenting all of these at once can be overwhelming and lead to a fractured or inaccurate mental model. Instead, the interface should initially present only the most essential features, making it easy for users to get started and build a foundational understanding. As users become more familiar with the system, or as they encounter specific needs, more advanced features and options can be revealed. This could be achieved through expandable menus, contextual help prompts, or guided tutorials that unlock as the user progresses. For example, a sophisticated AI-powered data analysis tool might initially present a simple interface for uploading data and requesting basic descriptive statistics. Only when the user attempts more complex operations, or actively seeks them out, would options for advanced statistical modeling, machine learning model training, or custom visualization tools be made available. This gradual introduction ensures that users build their mental model incrementally, focusing on core functionality before delving into more intricate aspects, thereby preventing cognitive overload and fostering a more robust and accurate understanding.

The language used in AI interfaces and prompts plays a significant role in shaping mental models. Employing clear, concise, and unambiguous language is crucial. Technical jargon should be avoided unless it is absolutely necessary and is accompanied by a clear explanation. Using terms that are familiar to the user's domain of expertise or general knowledge helps bridge the understanding gap. For instance, if an AI is designed to assist with project management, using terms like "tasks," "deadlines," "dependencies," and "milestones" aligns with standard project management terminology and aids in building an accurate mental model of the AI's role and capabilities within that context. Conversely, if the AI uses proprietary or overly abstract terms, users will struggle to map these to their existing knowledge, leading to confusion. The tone of the AI's communication is also important. An overly casual or overly formal tone, if mismatched with the user's expectations or the task at hand, can also contribute to a less accurate mental model. For example, an AI providing critical financial advice should maintain a professional and serious tone, whereas a playful AI companion might adopt a more lighthearted demeanor.

When designing AI systems that learn and adapt, it is essential to make this learning process visible and controllable. A "black box" AI, whose internal workings are entirely hidden, can lead to users developing superstitious or inaccurate mental models. They might attribute human-like intentionality, bias, or even sentience to the AI's actions without understanding the underlying data-driven processes. Therefore, designing for transparency is key. This involves providing mechanisms for users to:

View learned information: Show users what data or patterns the AI has learned about them or their environment.

Understand the basis of recommendations/decisions: Offer clear, albeit simplified, explanations for why the AI made a particular suggestion or took a specific action. For instance, an AI recommending a news article could indicate it's based on the user's reading history of similar topics or current trending news.

Correct or modify learned information: Allow users to explicitly correct inaccuracies or guide the AI's learning process. If an AI misinterprets a user's preference, the user should have a straightforward way to correct it, and see that correction take effect.

Consider an AI-powered music streaming service. If the AI starts recommending songs from a genre the user actively dislikes, and there's no obvious way to tell the AI "I don't like this genre," the user's mental model of the AI's intelligence degrades. They might feel the AI is "dumb" or "not listening." However, if the interface includes a "thumbs down" button on recommendations that allows the user to specify "not interested in this genre," or even a "block artist" feature, this provides a tangible way for the user to influence the AI's learning and update their mental model to understand that the AI is responsive to explicit feedback, even if its initial predictions were off.

The design of error messages and fallback mechanisms also significantly influences mental models. When an AI encounters an error or cannot fulfill a request, the message provided should not only state that an error occurred but also offer guidance on how to proceed. Instead of a generic "Error occurred," a more helpful message might be: "I couldn't understand your request. Please try rephrasing it, or ask for help." This guides the user towards actions that are likely to resolve the issue and helps them understand the AI's limitations. If the error is due to an external factor, such as a network issue or a system overload, the AI should ideally communicate this. "I'm having trouble connecting to the network right now. Please try again in a few minutes." This transparency helps the user form a mental model that includes external dependencies, rather than solely blaming the AI's internal logic.

The iterative nature of user interaction with AI systems necessitates a design approach that anticipates and supports ongoing learning and refinement of mental models. Users are not static entities; their understanding of a system evolves with each interaction. Therefore, AI systems should be designed to detect and respond to patterns of user behavior that indicate confusion or misunderstanding. For example, if a user repeatedly navigates back and forth between two screens or repeatedly attempts an action that fails, the AI could infer a potential gap in understanding. In such cases, the system could proactively offer contextual help, such as a pop-up tip, a brief tutorial on the relevant feature, or a link to a more detailed explanation. This active support from the AI helps guide the user towards a more accurate mental model without them having to explicitly seek assistance. This forms a positive feedback loop: the AI observes user behavior, infers potential misunderstanding, provides assistance, and the user's mental model is corrected and strengthened.

Ultimately, designing for intuitive AI understanding is an ongoing commitment to empathy and clarity. It requires designers to step into the user's shoes, anticipate their needs and potential points of confusion, and build systems that are not only powerful but also comprehensible. By leveraging familiar metaphors, maintaining consistent interaction patterns, providing clear affordances, and fostering transparency, we can empower users to build accurate mental models of AI systems. This, in turn, leads to more effective, efficient, and trustworthy human-AI collaboration, unlocking the true potential of artificial intelligence to augment human capabilities. It transforms AI from a mysterious, unpredictable force into a comprehensible, reliable, and valuable partner.

Building an accurate mental model of an artificial intelligence system hinges on a fundamental principle: users must be explicitly informed about what the AI can and, crucially, what it cannot do. This clarity is not merely a matter of good practice; it is foundational to preventing user frustration, mitigating misuse, and fostering genuine trust. When users are left to infer the boundaries of an AI's capabilities, they are prone to developing misconceptions that can lead to unmet expectations and a fractured understanding of the system's utility. Therefore, a proactive and transparent approach to communicating these capabilities and limitations is paramount.

The initial introduction to an AI system, often through an onboarding process, presents a critical opportunity to establish these boundaries. This is where users first encounter the AI and begin to form their initial mental models. Instead of allowing users to stumble upon limitations through trial and error, onboarding sequences can proactively highlight key functionalities and implicitly or explicitly signal areas where the AI falls short. For instance, a new user interacting with an AI-powered writing assistant might be greeted with a brief, interactive tutorial. This tutorial could showcase the AI's ability to suggest sentence completions, rephrase awkward phrasing, and check for grammatical errors. Simultaneously, it could subtly manage expectations by indicating that while the AI can assist with content generation, it is not designed to produce fully polished, publishable articles without significant human oversight. This upfront communication ensures that the user understands the AI as a powerful co-pilot rather than an autonomous author, thereby setting a realistic trajectory for their interaction.

Beyond the initial onboarding, ongoing communication through in-context cues and readily accessible help resources is vital for reinforcing and refining a user's understanding. Tooltips, for example, can provide concise explanations of specific features and their intended scope. Hovering over a particular button in an AI-driven data visualization tool might reveal a tooltip stating, "This feature generates basic charts and graphs from your uploaded data. For advanced statistical analysis, please use the 'Analysis Suite' option." This immediate, contextual information serves to clarify the affordances of the element the user is interacting with and gently guides them towards the appropriate functionalities for different tasks. It prevents the user from assuming that a feature designed for simple charting can somehow perform complex inferential statistics, thus avoiding a potential dead end and the subsequent frustration.

Comprehensive and easily discoverable help documentation acts as another cornerstone for communicating AI capabilities and limitations. This documentation should go beyond a mere list of features and delve into the practical implications of using the AI. For a system designed to analyze customer feedback, the help section could include specific examples of the types of sentiment it can accurately identify (e.g., positive, negative, neutral) and also clearly state the limitations. It might explain that nuanced emotions, sarcasm, or highly idiomatic expressions can sometimes be misinterpreted. Furthermore, it could outline the types of data the AI performs best with, such as structured survey responses, versus data it might struggle with, like informal social media comments laden with slang. By providing this detailed, yet accessible, information, users can calibrate their expectations and understand how to best leverage the AI's strengths while being mindful of its potential weaknesses. This informed approach allows users to tailor their input and interpretation of the AI's output, leading to more effective and accurate outcomes.

Consider an AI-powered image editing application. A user might upload a photograph and expect the AI to perform advanced graphic design tasks, such as adding intricate illustrations, creating complex layouts, or generating entirely new visual elements from scratch. However, if the AI's core functionality is focused on photo enhancement – adjusting brightness, contrast, color balance, or applying stylistic filters – then clearly delineating these capabilities is essential. This communication can occur through various design elements. The application's welcome screen or a "What's New" section could prominently feature a statement like: "Enhance your photos with intelligent adjustments, creative filters, and smart retouching. For professional graphic design, please use dedicated design software." Tooltips next to specific adjustment sliders could explain their function in simple terms, e.g., "Smart Contrast: Automatically balances light and dark areas for a more visually appealing image." Crucially, any attempt to perform a graphic design task beyond the AI's scope should be met not with a cryptic error message, but with a polite and informative response. For instance, if a user tries to drag and drop a vector illustration into the image editor, the AI could respond with a message such as, "This tool is designed for photo editing. To add graphic elements, please use our companion design application or export your image and import it into a graphic design program." This prevents the user from believing the AI is broken or incapable, instead guiding them towards the appropriate tools and fostering a more accurate understanding of the AI's intended purpose and boundaries.

The language used in system prompts and feedback mechanisms is a powerful, yet often understated, tool for communicating these boundaries. Instead of generic messages, prompts can be crafted to explicitly manage user expectations. For an AI that generates code snippets, a prompt might read: "Describe the function you want to create. I can help with common programming tasks, but complex algorithms or system-level integrations may require manual implementation." This immediately informs the user that the AI is a helper for specific, manageable coding tasks, not a replacement for a senior software architect. Similarly, when an AI provides output, the accompanying explanation or summary can also reinforce limitations. If an AI analyzes a lengthy document and produces a summary, it could add a disclaimer like, "This summary captures the main points based on keyword frequency and sentence structure. For in-depth thematic analysis, further review is recommended." This subtle but important addition helps the user understand that the AI's analysis is based on certain methodologies and has inherent limitations, preventing them from treating the summary as an exhaustive or perfectly nuanced interpretation.

Furthermore, the design of error states should be viewed not just as indicators of malfunction, but as opportunities for communication. A simple "Error" message is unhelpful. An AI system that fails to process a request should ideally provide context about *why* it failed and what the user can do next. If an AI recommender system cannot find suitable recommendations, instead of just showing a blank page, it could display a message like, "We couldn't find recommendations that perfectly match your current preferences and recent activity. Try broadening your search criteria or exploring some of our trending categories." This feedback not only acknowledges the system's failure to meet the request but also offers actionable advice, implicitly communicating the parameters that influence its recommendation engine and guiding the user towards successful interaction. This approach turns a potential point of failure into a learning experience, reinforcing the user's understanding of how the AI operates and what factors influence its performance.

Transparency in the learning process of an AI is also a critical component of communicating its limitations. While it's often impractical to expose the intricate details of machine learning algorithms, users should understand that the AI learns from data and can therefore exhibit biases present in that data, or make errors based on insufficient or misleading information. For instance, an AI designed to assist with job applications might be trained on historical hiring data.

If that historical data reflects societal biases, the AI might inadvertently perpetuate them in its recommendations or resume analysis. In such cases, it is incumbent upon the system designers to communicate this possibility. This could be done through an introductory statement within the application's privacy policy or a dedicated FAQ section, explaining that the AI is trained on past data and that users should critically evaluate its suggestions, especially in sensitive areas like employment. Providing users with options to report biased outputs or to manually override AI-driven decisions further empowers them and reinforces the understanding that the AI is a tool that requires human oversight, rather than an infallible arbiter.

The distinction between generative AI and discriminative AI, though technical, has practical implications for user understanding. A generative AI, like those that create text or images, often operates on principles of probability and pattern synthesis. Users might anthropomorphize these systems, believing they "understand" or "create" in a human sense. Communicating that these systems generate outputs based on statistical relationships in vast datasets, rather than genuine comprehension or intent, is crucial. For example, when a generative AI produces creative text, a subtle footer indicating "Generated using AI based on [mention general type of training data]" can help anchor the user's perception. Conversely, a discriminative AI, which classifies or predicts based on input (e.g., spam filters, image recognition), operates differently. Users need to understand that its accuracy is dependent on the quality and representativeness of the training data for classification tasks. If a spam filter starts misclassifying legitimate emails, it's not "maliciously" targeting the user; it's likely that the patterns of legitimate emails have changed, or the training data has become outdated, leading to a degradation in the AI's discriminatory power. Clearly communicating this dynamic allows users to understand the AI's performance not as a deliberate act, but as a consequence of its underlying operational principles and the data it has processed.

In essence, communicating AI capabilities and limitations is an ongoing dialogue. It begins with clear onboarding, is reinforced through intuitive interface design and accessible help resources, and is sustained by transparent feedback mechanisms and carefully managed error states. By proactively informing users about what an AI can and cannot do, designers enable the formation of accurate mental models. This, in turn, leads to more effective, less frustrating, and ultimately more trustworthy human-AI collaboration, allowing users to harness the full potential of these powerful tools without succumbing to the pitfalls of misunderstanding. It is through this commitment to clarity that we can ensure AI systems are not perceived as opaque or unpredictable entities, but as comprehensible and valuable assistants.

The journey of building an accurate mental model for users interacting with artificial intelligence is not a static one. It's a dynamic, iterative process, much like the refinement of the AI models themselves. While initial onboarding, in-context cues, and comprehensive documentation lay a robust foundation, the true depth of understanding is forged through continuous interaction and, critically, through mechanisms that allow users to actively participate in the learning and correction process. This is where the concept of feedback loops becomes not just beneficial, but essential. These loops are designed to capture the user's experience, their moments of clarity, and, perhaps more importantly, their moments of confusion or disagreement, and then channel that information back into refining both the user's mental model and the AI's operational parameters.

Consider the nature of many AI systems today: they are not simply passive tools but interactive agents. This interactivity inherently implies a dialogue, and a dialogue requires more than just unidirectional instruction. Users need avenues to express their understanding, to challenge the AI's assumptions, and to guide its behavior. When an AI makes a mistake, or when its output is misinterpreted by the user, a well-designed feedback loop provides a pathway for correction. This correction can operate on two distinct but interconnected levels. First, it allows the user to rectify their own evolving mental model. If the AI misunderstands a command or provides an output that is demonstrably incorrect, the user's ability to provide corrective feedback helps them recalibrate their expectations and better understand the AI's limitations or specific operational quirks. For instance, if a user instructs an AI scheduling assistant to book a meeting for "tomorrow morning" and the AI interprets this as the immediate morning of the current day, rather than the following calendar day, the user's correction – perhaps by specifying "next Tuesday morning" – serves to highlight the ambiguity of their initial request and the AI's interpretation. This interaction reinforces the understanding that the AI relies on explicit and unambiguous language.

Second, and equally crucial, these feedback loops provide invaluable data for the developers to refine the AI model itself. Every correction, every piece of user feedback, is a data point. When a user flags an incorrect output, they are not just expressing dissatisfaction; they are inadvertently contributing to the AI's training set, highlighting areas where its algorithms might be underperforming or misaligned with user intent. For an AI-powered content moderation system, for instance, users who are tasked with reviewing flagged content might have the ability to override the AI's decision. If the AI incorrectly flags a benign post as harmful, a human reviewer can mark it as "false positive." This feedback directly informs the model, helping it learn to distinguish more accurately between genuine violations and acceptable content in the future. The system, in turn, can then provide better, more accurate results, leading to a more positive user experience and a more reliable AI.

Designing these feedback mechanisms requires careful

consideration of the user's workflow and cognitive load. They should be intuitive, easily accessible, and, whenever possible, integrated seamlessly into the interaction flow. A common and effective approach is to prompt users for feedback at points where uncertainty or potential misunderstanding is high. For example, a smart assistant, after performing a complex task or generating a nuanced piece of content, might follow up with a question like, "Did this response fully address your needs?" or "Is there anything unclear about the information provided?" The options for response should be simple: a quick "Yes" or "No," a rating scale, or a brief option to provide more detail. If the user indicates dissatisfaction or confusion, the system can then offer more specific prompts, such as "What was unclear?" or "How could this be improved?" This layered approach ensures that feedback is collected efficiently for most interactions while allowing for deeper qualitative insights when needed.

Consider the scenario of an AI-driven diagnostic tool for medical professionals. While the AI can analyze scans and suggest potential diagnoses, it's imperative that the human physician remains in control and can validate or refute the AI's findings. The interface could incorporate clear buttons or options within the AI's output display, allowing the physician to easily mark a suggestion as "Confirmed," "Unlikely," or "Needs Further Investigation." Each of these actions provides a direct signal to the AI and its developers. A "Confirmed" might strengthen the AI's confidence in that particular diagnostic pathway for similar future cases. An

"Unlikely" or "Needs Further Investigation" flag, especially if aggregated across multiple users and cases, indicates a need for the AI to re-evaluate its underlying model, perhaps by seeking additional data or adjusting its confidence thresholds for certain indicators. This iterative process of physician validation and AI refinement is critical for building trust and ensuring the AI's safety and efficacy in a high-stakes domain.

Another powerful form of feedback loop involves proactively identifying and addressing user confusion. AI systems can be designed to detect signals of uncertainty in user behavior. This might include repeated attempts to rephrase a query, hesitant cursor movements, or a series of actions that deviate from typical usage patterns. When such signals are detected, the AI can interject not with an error message, but with a clarifying question or suggestion. For instance, a natural language processing interface might notice a user struggling to articulate a complex request. Instead of waiting for the user to abandon the interaction, the AI could offer a prompt such as, "It seems you're trying to achieve X, but I'm not sure I fully understand. Would you like me to try X, Y, or Z?" or "Perhaps phrasing it as [suggested alternative phrasing] would help me process your request more effectively." This not only helps the user articulate their intent more clearly, leading to a more successful outcome, but also implicitly teaches the user about the types of phrasing and structure the AI understands best, thereby refining their mental model of the system's input requirements.

These proactive interventions also serve to mitigate the frustration that often arises from a mismatch between user expectation and AI capability. By anticipating potential confusion, the AI can guide the user towards a more productive interaction path. This is particularly relevant for generative AI systems, where the creative output can sometimes be unexpected or diverge significantly from what the user envisioned. If an AI art generator produces an image that is artistically distant from the user's prompt, instead of simply presenting the result, it could offer options like "Regenerate with these adjustments," "Focus more on [specific element]," or "Try a different style." These options act as granular feedback mechanisms, allowing the user to steer the generation process in real-time and learn about how the AI interprets stylistic prompts, color palettes, and compositional elements. The user's choices in navigating these options become valuable training data for the AI, helping it to better align its creative outputs with user intent in subsequent interactions.

The design of these feedback mechanisms should also consider different levels of user engagement. Not all users will have the time or inclination to provide detailed feedback. Therefore, a tiered approach is often most effective. At the simplest level, implicit feedback can be gathered by observing user behavior. If a user consistently ignores AI-generated suggestions or bypasses certain AI-powered features, this can be interpreted as a signal of disinterest or dissatisfaction. On a more explicit level, simple binary feedback (e.g.,

"Was this helpful? Yes/No") can be collected with minimal user effort. For users who are more engaged or for critical scenarios, providing an option for free-text feedback or a more detailed questionnaire allows for richer qualitative data. This data can then be analyzed to identify patterns, understand common pain points, and pinpoint specific areas for model improvement.

Furthermore, the concept of "correction" can be extended beyond simply flagging errors. In collaborative AI systems, users might need to actively correct the AI's understanding of shared context or evolving goals. Imagine a project management AI that tracks tasks and deadlines. If a team member makes a significant change to a project timeline that is not immediately communicated to the AI, the AI might continue to operate based on outdated information. In such cases, providing a mechanism for users to explicitly update the AI's understanding of project status, priorities, or dependencies is crucial. This could involve a simple "Update Status" button linked to a form where users can specify changes, or even a natural language interface where users can say, "The deadline for task X has been moved to next Friday." This direct input corrects the AI's internal representation of the project and ensures its subsequent advice and notifications are aligned with the current reality, thereby refining the user's mental model of the AI as a reliable project partner rather than an entity operating in a vacuum.

The feedback loop also plays a critical role in building user confidence and fostering a sense of agency. When users know that their feedback is heard and acted upon, they are more likely to engage with the AI system positively. This is especially true for systems that are still under development or have known limitations. By providing clear channels for reporting issues or suggesting improvements, designers demonstrate a commitment to user experience and continuous improvement. This can transform a potentially negative experience – encountering an AI error – into a constructive one where the user feels like a valued participant in the system's evolution. The AI then becomes less of an opaque black box and more of a collaborative tool, responsive to user input.

Moreover, the feedback loop can be leveraged to help users understand the underlying principles of the AI without necessarily diving into technical jargon. When a user provides feedback, the system can respond not just by acknowledging the input but by offering a brief, user-friendly explanation of why the AI might have behaved in a certain way and how the feedback will be used. For example, if a user flags an AI-generated summary as missing a key point, the system could respond with something like, "Thank you for your feedback. Our AI currently prioritizes information based on frequency of mention and explicit topic headings. We're working on improving its ability to detect more nuanced themes." This type of explanation, embedded within the feedback response, helps the user understand the AI's current methodology and its developmental trajectory, further solidifying their mental model. It demystifies the AI's operations and sets realistic expectations for its future capabilities.

The continuous cycle of user feedback, AI refinement, and subsequent user adaptation is what ultimately leads to the formation of accurate, robust, and adaptive mental models. It's a testament to the idea that building effective AI is not just about algorithmic prowess, but about creating a symbiotic relationship where both the human and the machine learn from each other. These feedback mechanisms are the connective tissue that allows this learning to occur, turning potential points of friction into opportunities for deeper understanding and more powerful collaboration. Without them, even the most sophisticated AI risks becoming a static entity, its potential limited by the user's static or incorrect assumptions about its capabilities. Therefore, investing in well-engineered feedback loops is not an optional add-on; it is an indispensable component of designing AI systems that are truly effective, trustworthy, and user-centric.

In the realm of complex decision support, where the stakes are high and the potential for error carries significant consequences, the development of accurate user mental models is paramount. Consider a hypothetical AI system designed to assist financial analysts in portfolio management. This system, let's call it "Predictive Capital," ingests vast amounts of market data, economic indicators, company financial statements, and news sentiment to generate investment recommendations and risk assessments. The challenge for the design team was not merely to build a powerful analytical engine, but to ensure that the human analysts wielding this tool could form a reliable understanding of how it arrived at its conclusions, what its limitations were, and when to trust its output versus when to exercise their own professional judgment.

The interface of Predictive Capital was meticulously crafted to facilitate this understanding. Rather than presenting a final recommendation as a black box, the system offered a layered approach to its analysis. When Predictive Capital flagged a particular stock for potential underperformance, the initial output wasn't just a simple "sell" or "avoid" alert. Instead, it presented a concise summary of the key factors driving the recommendation, such as "increasing debt-to-equity ratio," "declining revenue growth," and "negative sentiment in recent earnings calls." Crucially, each of these summarized factors was hyperlinked. Clicking on "increasing debt-to-equity ratio," for instance, would not only provide the specific numerical value and trendline for that metric but also offer contextual information: what constitutes a "healthy" ratio in that industry, historical benchmarks, and how this specific company's ratio compares to its peers. This granular drill-down capability was a deliberate design choice to expose the underlying data and reasoning, allowing analysts to probe the evidence themselves.

This approach directly addressed the need for users to

build an understanding of the AI's analytical processes. An analyst seeing the debt-to-equity ratio flagged might have a pre-existing mental model that this metric is only a minor concern for companies in that particular sector. By being able to immediately access comparative data and industry norms, they could either validate the AI's concern, perhaps realizing their initial assumption was outdated or incomplete, or they could dismiss the AI's flag based on their own expert knowledge and the provided context. This interaction transforms the AI from a prescriptive oracle into a sophisticated analytical assistant, presenting its findings as well-supported hypotheses rather than immutable truths. The mental model being built here is not just about *what* the AI recommends, but *why* it recommends it, and crucially, *how* reliable that "why" is in relation to the analyst's own expertise.

Furthermore, Predictive Capital incorporated an "explainability layer" that went beyond simply listing data points. When the AI identified a particular news sentiment as a contributing factor to its recommendation, the system would display snippets of the actual news articles and social media posts that influenced its sentiment analysis. This allowed analysts to assess the quality and relevance of the source material and the AI's interpretation of it. For example, if the AI flagged a piece of social media chatter as negative sentiment, the analyst could read the post and decide if the AI's classification was appropriate. Perhaps the post was sarcastic, or its context was misunderstood. This direct exposure to the source data and the AI's processing of it is vital. It helps the user understand the AI's capabilities and limitations in interpreting nuanced human language, which is notoriously difficult for machines. This transparency fosters a more robust mental model, where the user appreciates that the AI's sentiment analysis is a sophisticated algorithm but still susceptible to misinterpretation, especially with idiomatic expressions or evolving slang.

The system also included a feedback mechanism specifically designed to refine user mental models and, in turn, improve the AI. Analysts were provided with options to challenge the AI's conclusions or to affirm them with additional insights. If an analyst disagreed with the AI's sentiment analysis, they could flag it and provide a brief explanation, perhaps by selecting the correct sentiment category or highlighting a misleading phrase. This feedback loop served a dual purpose. Firstly, it allowed the analyst to mentally correct their interpretation of the AI's capabilities. If they initially believed the AI was infallible in sentiment analysis, a series of incorrect flags would quickly adjust that mental model. Secondly, this aggregated feedback data was used to retrain the AI's natural language processing modules, making it more accurate over time. This continuous refinement process is key in high-stakes domains; the AI doesn't just learn, but its users also learn about the AI, and this co-learning solidifies trust and efficacy.

Another critical aspect of Predictive Capital's design

was its handling of uncertainty and confidence levels. The AI didn't present all recommendations with the same degree of certainty. For each recommendation, it provided a confidence score, visually represented through a spectrum or a numerical value. More importantly, the system's interface was designed to draw the user's attention to recommendations with lower confidence levels. These might be accompanied by prompts such as, "This recommendation has a moderate confidence level due to conflicting indicators. Further human review is advised," or "Confidence score is low. Consider alternative scenarios." This proactive communication of uncertainty is a cornerstone of effective decision support. It prevents users from blindly following a recommendation that the AI itself deems less reliable. The analyst's mental model is thus shaped to understand that the AI is not always definitive, and that there are degrees of risk and confidence associated with its outputs. This encourages a more critical and informed approach to decision-making.

Consider a medical diagnosis AI, a field where similar principles of decision support and mental model accuracy are even more critical. An AI designed to assist radiologists in identifying cancerous nodules in medical scans would need to be equally transparent. When the AI flags a potential nodule, it shouldn't just highlight an area on the scan. It should provide an explainable report detailing
why it believes it's a nodule, referencing specific visual characteristics—shape, density, texture, margin

characteristics—and potentially comparing it to a database of known benign and malignant growths. The interface might allow the radiologist to click on these characteristics and see examples from the database, or even see how the AI's classification algorithm weighed each feature. If the AI flags a lesion that the radiologist believes is benign, they should be able to easily contest the flag, perhaps by indicating specific benign characteristics that the AI might have overlooked or misinterpreted. This direct engagement with the AI's reasoning process is crucial for building trust. A radiologist needs to understand if the AI is missing subtle signs, overemphasizing certain features, or suffering from image artifacts. This understanding directly shapes their mental model of the AI as a reliable second opinion rather than a replacement for their expertise.

The integration of data provenance information is also a powerful tool for building accurate mental models in complex decision support systems. In the financial domain, for example, if Predictive Capital recommends a particular investment based on a news article, the system should clearly state the source, the date of publication, and potentially the author or publication's historical reliability score. If the recommendation is based on a specific financial statement, it should link directly to the relevant section of that statement. This traceability allows the user to verify the information and understand the basis of the AI's claims. It helps to demystify the AI's process and builds confidence by demonstrating that its recommendations are grounded in verifiable data. Users learn to trust not just the AI's output, but the integrity of the data sources it utilizes and the logic it applies to them.

The design of interactive visualizations also plays a crucial role. Instead of static charts, Predictive Capital might employ dynamic visualizations that allow analysts to manipulate parameters and see how the AI's recommendations change in real-time. For instance, an analyst could adjust the weighting of different risk factors—economic stability, industry trends, company-specific performance—and observe how the AI's portfolio allocation suggestions adapt. This hands-on exploration allows users to intuitively grasp the relationships between different variables and the AI's decision-making process. It moves beyond a passive reception of information to an active exploration, which is far more effective in forging deep and accurate mental models. The user learns by doing, by experimenting, and by observing the AI's responses to their interventions. This active engagement is more likely to lead to a mental model that is not only accurate but also flexible and adaptive to new information or altered circumstances.

In essence, the success of AI in complex decision support hinges on a symbiotic relationship between the human expert and the artificial intelligence. This relationship is built on a foundation of trust, which in turn is forged through transparency, explainability, and the continuous refinement of user understanding. By designing systems that allow users to probe, question, and interact with the AI's reasoning, we empower them to develop accurate mental models. These models enable users to leverage the AI's analytical power effectively, to identify its limitations, and ultimately, to make more informed and confident decisions in high-stakes environments. The case of Predictive Capital, and similar systems in fields like medicine, demonstrates that the most powerful AI is not necessarily the most opaque, but rather the one that most effectively cultivates understanding and collaboration with its human counterpart. The goal is not to replace human judgment, but to augment it, and this augmentation is only truly effective when the human user possesses an accurate and nuanced mental model of the AI they are using. This iterative process of interaction, explanation, and feedback is the engine that drives both user proficiency and AI improvement, leading to more robust and reliable decision-making.

Chapter 8: Essential AI Interaction Patterns

Conversational interfaces, encompassing chatbots and voice assistants, represent a significant evolution in how humans interact with technology. Moving beyond graphical user interfaces (GUIs) and command-line interactions, these systems leverage natural language to facilitate communication. The underlying challenge in designing effective conversational AI lies in replicating the nuances and complexities of human conversation, a feat that involves deep understanding of linguistics, user psychology, and sophisticated computational linguistics. The goal is to create interactions that feel natural, intuitive, and efficient, whether through text-based chat or spoken dialogue. This requires careful consideration of numerous design principles, from the initial understanding of user intent to the final delivery of a helpful response.

At the heart of any conversational interface is its ability to understand what the user is saying or typing. This capability relies heavily on Natural Language Understanding (NLU), a subfield of artificial intelligence focused on enabling machines to comprehend human language. NLU systems parse sentences, identify entities (like names, dates, locations), determine the intent behind the utterance (e.g., "book a flight," "check the weather," "play music"), and extract relevant parameters. The sophistication of NLU directly impacts the user experience. A system with weak NLU might misunderstand simple requests, leading to frustration. For instance, a user asking, "I want to book a train ticket from London to Paris for tomorrow," might be met with confusion if the NLU system fails to correctly identify

"London" and "Paris" as departure and arrival cities, or misinterprets "tomorrow" as a specific date.

The challenges in NLU are manifold. Human language is rife with ambiguity, colloquialisms, slang, sarcasm, and context-dependent meanings. Consider the word "bank." In one context, it refers to a financial institution; in another, it means the side of a river. A robust NLU system must be able to disambiguate such words based on the surrounding dialogue. Similarly, phrasal verbs and idiomatic expressions pose significant hurdles. A phrase like "kick the bucket" has a meaning far removed from its literal interpretation. Voice assistants face additional challenges, including variations in accents, background noise, and the nuances of prosody (intonation, rhythm, and stress), which can alter the meaning of an utterance. Designing NLU models that can accurately handle this linguistic variability requires vast amounts of training data and sophisticated machine learning techniques, often involving deep learning models like recurrent neural networks (RNNs) and transformers.

Once the user's intent is understood, the conversational AI must engage in dialogue management. This is the process of orchestrating the conversation, deciding what the system should say or do next, and maintaining the context of the interaction. Effective dialogue management ensures that the conversation flows logically and cooperatively. A key aspect of this is managing conversational turns and ensuring smooth transitions between topics. For instance, if a user asks about the weather and then immediately asks for a restaurant recommendation, the system needs to recognize the shift in topic and respond appropriately without losing track of the initial weather inquiry if it was part of a larger goal.

A crucial element of dialogue management is the ability to ask clarifying questions. When the NLU system encounters ambiguity or insufficient information, it should prompt the user for more details rather than making assumptions or failing outright. This pattern, known as a clarifying dialogue, is essential for robust error recovery and for guiding the user towards a successful outcome. For example, if a user says, "Book me a hotel," the system should not guess a location or dates. Instead, it should ask, "Where would you like to book a hotel?" or "For what dates would you like to book a hotel?" These clarifying questions demonstrate that the system is trying to understand and assist, fostering a sense of helpfulness. The design of these clarifying questions is critical; they should be concise, clear, and lead the user towards providing the necessary information.

Error recovery dialogues are another vital component of conversational AI. No matter how sophisticated the NLU and dialogue management, errors will occur. The system needs to gracefully handle situations where it misunderstands the user, or the user provides information that cannot be processed. A well-designed error recovery dialogue can salvage an interaction that might otherwise be abandoned. This might involve apologizing for the misunderstanding, rephrasing the question, or offering alternative interpretations. For instance, if a voice assistant mishears a command, it could respond with, "I'm sorry, I didn't quite catch that. Did you say 'set a timer for five minutes'?" This not only highlights the error but also offers a potential correction, allowing the user to confirm or deny. More complex error recovery might involve guiding the user through a predefined troubleshooting process if a command consistently fails.

Providing clear feedback to the user is paramount in conversational interfaces. Users need to know that the system has understood them and what action it is taking. This feedback can be explicit, such as a confirmation message ("Okay, I've added milk to your shopping list"), or implicit, like a visual cue in a chatbot indicating that the system is "typing" or processing a request. In voice assistants, auditory cues, such as a gentle chime or a specific sound, can indicate that the device is listening or has processed a command. Without clear feedback, users are left in a state of uncertainty, unsure if their input was registered or if the system is functioning correctly. This uncertainty can quickly lead to frustration and disengagement.

Proactive engagement is a more advanced interaction pattern that can significantly enhance the user experience. Instead of merely reacting to user commands, proactive conversational agents can anticipate user needs and offer assistance or relevant information. For example, a travel assistant might proactively inform a user about a flight delay or suggest alternative routes due to traffic. A smart home assistant could remind a user to turn off the lights before they leave the house, based on their typical routine. This requires the AI to learn user patterns and preferences, and to have access to relevant real-world data. When executed thoughtfully, proactive engagement can make the AI feel more intelligent and indispensable, transforming it from a passive tool into an active assistant. However, it must be implemented judiciously to avoid being intrusive or annoying. The system should provide users with control over the level of proactivity they desire.

The design of the conversational persona is also a critical aspect. Should the chatbot be formal or informal? Humorous or serious? Empathetic or purely functional? The persona should align with the brand and the intended use case. A customer service chatbot for a bank might adopt a professional and reassuring tone, while a chatbot for a gaming app might be more playful and engaging. The persona is conveyed through word choice, sentence structure, and even the use of emojis or animated avatars in text-based interfaces. Consistency in persona is key to building a coherent and trustworthy interaction. A chatbot that switches between being overly casual and sternly professional can be jarring for the user.

Designing for different communication modalities – text versus voice – presents unique considerations. Text-based chatbots can leverage visual elements like buttons, carousels, and rich media to present information and guide users. They also allow users to review the conversation history easily. Voice assistants, on the other hand, must rely solely on audio cues. This means that responses need to be concise, clear, and well-paced. Complex information, such as a list of options, might be better presented visually in a companion app or by providing a summary followed by an offer to elaborate. The cognitive load on the user is also different; listening requires continuous attention, while reading allows for more flexibility in processing information.

Consider the application of these principles in various scenarios. In customer service, chatbots are increasingly used to handle frequently asked questions, troubleshoot common issues, and guide users through processes like account setup or order tracking. A well-designed customer service bot can reduce wait times, provide 24/7 support, and free up human agents for more complex queries. The dialogue might start with a broad question like, "How can I help you today?" followed by a menu of common options or an invitation to type a query. If the bot cannot resolve the issue, it should have a seamless handover mechanism to a human agent, providing the agent with the conversation history.

In e-commerce, chatbots can act as personal shoppers, recommending products based on user preferences, answering questions about product features, and assisting with the checkout process. For instance, a user might say, "I'm looking for a new pair of running shoes, size 10, for trail running." The chatbot would then need to access product catalogs, filter by the specified criteria, and present suitable options, perhaps with images and brief descriptions. It might then ask follow-up questions like, "Are you looking for waterproof shoes?" or "Do you have a preferred brand?"

Smart home systems, powered by voice assistants like Amazon Alexa or Google Assistant, exemplify the integration of conversational AI into daily life. Users can control lights, adjust thermostats, play music, set reminders, and get news updates using voice commands. The design here focuses on ease of use and natural language understanding for a wide range of commands. For example, a user might say, "Hey Google, turn down the living room lights to 50%," or "Alexa, play some jazz music." The system needs to parse these commands, identify the device or service to control, and execute the action. Error handling in this context is crucial; if the assistant misunderstands, it could lead to unwanted actions, like turning off the wrong lights.

The development of conversational AI is an iterative process. Designers and engineers continuously collect data on user interactions, analyze where conversations break down, and use this information to refine the NLU models, dialogue flows, and persona. User testing is essential, observing how real users interact with the system and identifying pain points. Metrics like task completion rate, user satisfaction, and conversation length can help gauge the effectiveness of the design.

Furthermore, ethical considerations are paramount in the design of conversational interfaces. Transparency about the AI's capabilities and limitations is crucial. Users should be aware that they are interacting with a machine, not a human. Privacy concerns are also significant, especially when the AI collects personal data. Systems must be designed to protect user data and to be transparent about how it is used. The potential for bias in NLU models, trained on data that reflects societal biases, must also be addressed to ensure fair and equitable interactions. For example, an AI that consistently misunderstands commands from users with certain accents or dialects would be discriminatory.

In summary, creating effective conversational interfaces involves a multifaceted approach that goes beyond simply recognizing words. It requires sophisticated NLU to understand intent, intelligent dialogue management to guide the conversation, robust error handling to recover from misunderstandings, clear feedback to keep the user informed, and a well-defined persona to create a consistent and engaging experience. By mastering these interaction patterns, designers can build chatbots and voice assistants that are not only functional but also intuitive, helpful, and ultimately, delightful to use, seamlessly integrating technology into the fabric of our daily lives through the power of natural language. The ongoing evolution of AI technologies, particularly in areas like sentiment analysis and commonsense reasoning, promises to make these conversational experiences even richer and more human-like in the future.

Recommendation systems have become an indispensable part of our digital lives, subtly guiding our choices across a vast landscape of content and products. Think of the last time you browsed Netflix and were presented with a curated list of "Because you watched X," or scrolled through Amazon's "Customers who bought this item also bought..." section, or opened Spotify to find a personalized "Discover Weekly" playlist. These are not random suggestions; they are the output of sophisticated artificial intelligence designed to understand our preferences and introduce us to things we might enjoy, often before we even realize we're looking for them. The core challenge in designing these systems lies in bridging the gap between a user's implicit or explicit tastes and the immense, ever-growing catalog of available items, be they movies, books, music, or merchandise.

At the heart of most recommendation systems are algorithms that attempt to predict user interest. These algorithms can be broadly categorized into a few key approaches, each with its own strengths and weaknesses. The most prevalent of these is

collaborative filtering. This method operates on the principle that users who agreed in the past tend to agree in the future. In essence, it leverages the collective wisdom of the user base. If User A and User B have both liked or purchased many of the same items, and User A has recently expressed interest in a new item, collaborative filtering will infer that User B might also be interested in that same item. This is the engine behind many "people like you also liked" features. For instance, if you've enjoyed a particular science fiction novel, and many other readers who enjoyed that same novel also enjoyed another specific fantasy novel, the system might recommend that fantasy novel to you, even though its genre is slightly different. The power of collaborative filtering lies in its ability to uncover unexpected connections and recommend items based on broader community behavior, rather than just the inherent characteristics of the item itself. However, it suffers from the "cold-start problem": it struggles to make recommendations for new users with no history or for new items that haven't been interacted with by many users yet. It also can lead to a "popularity bias," where more popular items get recommended more frequently, potentially overshadowing niche or lesser-known gems.

Another fundamental approach is **content-based filtering**. This method, in contrast to collaborative filtering, focuses on the attributes or characteristics of the items themselves and the user's past interactions with those attributes. If a user has consistently rated science fiction movies highly, a content-based system would analyze the features of those movies (e.g., genre, actors, directors, plot keywords) and then recommend other movies that share similar features. For example, if you frequently watch films starring a particular actor or directed by a specific individual, a content-based system will identify this pattern and suggest other works by the same individuals. Similarly, if you listen to a lot of jazz music, the system will analyze the tempo, instrumentation, and artist tags of that music and recommend other jazz pieces with similar characteristics. The advantage of content-based filtering is that it can recommend items to users with unique tastes without relying on a large community of users. It's also effective for new items, provided their content can be described and analyzed. The primary limitation is that it can lead to a "filter bubble" or "echo chamber" effect, where users are only exposed to content very similar to what they already know, limiting the possibility of discovering entirely new genres or styles. It requires a rich and accurate description of the items, which can be a challenge for certain types of content, like abstract art or complex pieces of music where descriptive metadata might be insufficient.

More advanced recommendation systems often combine these two primary approaches, or incorporate other techniques.

Hybrid systems aim to leverage the strengths of different algorithms while mitigating their weaknesses. For instance, a hybrid system might use content-based filtering to make initial recommendations for a new user and then gradually incorporate collaborative filtering as more user interaction data becomes available. Another common hybrid strategy is **model-based collaborative filtering**, which uses machine learning techniques like matrix factorization (e.g., Singular Value Decomposition or Alternating Least Squares) to uncover latent factors that explain user-item interactions. These latent factors can represent abstract concepts like "preference for action movies" or "appreciation for intricate plotlines" that are not explicitly defined in the item's metadata.

Beyond the algorithmic core, the **presentation of recommendations** is critical for user engagement and satisfaction. Simply listing a set of recommended items is rarely sufficient. Effective design involves making the recommendations discoverable, understandable, and actionable. One of the most important design patterns here is providing **explanations** for *why* an item is being recommended. This transparency builds trust and helps users understand the system's logic, making them more likely to accept and act on the suggestions. Explanations can take various forms: "Because you watched 'Stranger Things'," "Customers who bought this laptop also bought this mouse," "Because you have listened to Taylor Swift," or "This item is trending in your area." These explanations serve multiple purposes. They validate the user's past behavior, demonstrating that the system has learned from their actions. They also provide context, helping the user assess the relevance of the recommendation. For a user who enjoyed a specific, dark fantasy novel, being recommended another with the explanation "similar dark themes and complex character arcs" is far more compelling than simply seeing it appear on a list. This explanatory layer transforms the recommendation from a passive suggestion into an informed insight.

The design should also empower users to **refine their preferences**. Recommendation systems are not static; they should evolve with the user. Allowing users to explicitly signal their likes and dislikes is crucial. This can be achieved through various mechanisms: star ratings, thumbs up/down buttons, "not interested" options, or even more granular feedback like "show me more like this" or "show me less like this." For example, on a music streaming service, if a user repeatedly skips a particular artist or song that the system has recommended, the system should learn from this negative feedback and adjust future recommendations accordingly. Conversely, if a user actively seeks out and listens to a specific genre or artist, that explicit signal should heavily influence future suggestions. This active participation makes the user feel more in control of their experience and leads to increasingly accurate and personalized recommendations over time. It also provides valuable data for the system to improve its models. Some platforms go further by allowing users to curate their own lists or even directly influence algorithmic parameters, such as adjusting sliders for "discovery" versus "familiarity."

A significant design challenge in recommendation systems is managing the inherent tension between **serendipity and relevance**. Relevance is paramount; users generally want recommendations that align with their existing interests. However, an overemphasis on strict relevance can lead to the aforementioned filter bubble, where users are trapped in a cycle of consuming content that is too similar to what they already know. Serendipity, on the other hand, refers to the delightful discovery of unexpected but enjoyable items. A truly great recommendation system should strike a balance, offering a stream of relevant content while also occasionally introducing users to novel experiences that broaden their horizons. Designing for serendipity involves introducing an element of controlled randomness or exploring less obvious connections in the data. For instance, a music recommendation system might notice that a user enjoys both indie rock and classical music, and then serendipitously recommend a contemporary classical piece that incorporates elements of rock instrumentation or melodic structures. The explanation for such a recommendation might highlight these crossover elements, making the unexpected suggestion more palatable. Achieving this balance requires careful tuning of algorithms and often involves exploring items that are adjacent to, rather than directly similar to, a user's core preferences.

Furthermore, the
context in which a recommendation is made plays a vital role. A user's needs and preferences can change depending on the time of day, their current activity, or even their mood. A system that understands this context can offer more timely and appropriate suggestions. For example, a streaming service might recommend calming music in the evening or upbeat workout playlists in the morning. A news

recommendation system could prioritize breaking news during the day and in-depth articles for leisure reading. This contextual awareness often relies on inferring user state from various data points, such as time of day, device usage patterns, or even explicit user input about their current goals.

The ethical considerations surrounding recommendation systems are also increasingly important.
Bias in data can lead to biased recommendations, perpetuating societal stereotypes. For example, if a system is trained on historical data where certain job roles were predominantly held by one gender, it might inadvertently recommend those roles more often to users of that gender, even if they are exploring unrelated career paths. Similarly, if training data is skewed towards a particular demographic, recommendations might be less relevant or even offensive to users from underrepresented groups. Designers must be vigilant in identifying and mitigating these biases through diverse datasets, fairness-aware algorithms, and rigorous testing.

Transparency and explainability are not just good for user experience; they are also ethical imperatives. Users have a right to understand why they are being shown certain content or products, especially when those recommendations influence significant decisions. The "black box" nature of some complex AI models can be problematic. Efforts towards "explainable AI" (XAI) are crucial in this domain to demystify the recommendation process and allow for accountability.

The
reinforcement of echo chambers and filter bubbles
is another significant ethical concern. While personalized recommendations can be convenient, they can also limit exposure to diverse perspectives, reinforcing existing beliefs and potentially contributing to societal polarization. Designers have a responsibility to consider mechanisms that promote exposure to a broader range of viewpoints and information, even if those items are outside a user's immediate inferred preferences. This might involve deliberately introducing items from different categories or offering users the option to explicitly seek out diverse content.

Moreover, the
persuasive power of recommendation systems, especially in e-commerce and social media, raises concerns about manipulation. Systems designed to maximize engagement or conversion rates can inadvertently exploit psychological vulnerabilities, encouraging impulse purchases or excessive consumption of certain types of content. Designing ethically requires prioritizing user well-being and autonomy over purely commercial or engagement-driven metrics. This involves setting clear boundaries on the persuasive power of recommendations and giving users robust control over their experience.

Finally, the collection and use of

user data for personalization raise significant privacy concerns. Recommendation systems often require vast amounts of data about user behavior, preferences, and even personal demographics. Ensuring that this data is collected with informed consent, stored securely, and used responsibly is paramount. Users should have a clear understanding of what data is being collected, how it is being used to generate recommendations, and they should have control over their data, including the ability to opt-out of certain types of personalization or to have their data deleted. The General Data Protection Regulation (GDPR) and similar privacy frameworks are setting important standards in this area, emphasizing principles like data minimization and purpose limitation.

In conclusion, recommendation systems are a powerful manifestation of AI's ability to personalize digital experiences. They leverage sophisticated algorithms like collaborative filtering, content-based filtering, and hybrid approaches to predict user preferences and introduce them to relevant items. However, the effectiveness of these systems hinges not only on the underlying algorithms but also on thoughtful design. Providing clear explanations for recommendations, enabling users to refine their preferences, balancing serendipity with relevance, and considering the context of the user are all crucial for creating engaging and trustworthy experiences. As these systems become even more sophisticated, addressing the ethical implications related to bias, transparency, echo chambers, persuasive power, and user privacy will be paramount to ensuring they serve users beneficially and responsibly, fostering discovery rather than constraint, and empowering informed choices rather than subtle manipulation.

The evolution of human-computer interaction has moved beyond mere responsiveness to a more anticipatory paradigm, where intelligent agents are designed not just to react to explicit commands, but to actively predict and fulfill user needs. This shift heralds a new era of proactive assistance, fundamentally altering how we engage with technology. Instead of a user meticulously managing every detail, AI agents can shoulder a significant burden, operating in the background to streamline workflows, prevent potential issues, and enhance overall efficiency. The core principle is to imbue these agents with an understanding of user goals, context, and preferences, enabling them to act on their behalf in a way that is both helpful and unobtrusive.

Imagine an AI assistant that doesn't wait for you to remember to book a follow-up meeting after a client call, but instead, based on the conversation's duration and content, proactively suggests optimal times to schedule the next engagement, even checking the availability of all participants and presenting a few concrete options directly within your calendar. This is the essence of proactive assistance: identifying a future need or task and taking preparatory or direct action before the user even has to articulate it. This could manifest in numerous ways, from managing complex scheduling logistics to automatically adjusting environmental settings in a smart home based on predicted occupancy and user preferences. The underlying intelligence is designed to infer intent and anticipate requirements from a confluence of signals, including past behavior, calendar entries, communication patterns, and sensor data.

A key challenge in developing these proactive agents lies in the delicate balance between helpfulness and intrusiveness. An agent that is too eager to assist, intervening too frequently or in ways that disrupt the user's current focus, can quickly become an annoyance rather than an aid. The goal is to achieve a state of "graceful intervention," where the agent's actions are perceived as timely, relevant, and genuinely beneficial, rather than disruptive or overwhelming. This necessitates a sophisticated understanding of user context. For instance, an agent might detect that a user is deep in a complex coding session and therefore defer non-critical notifications or suggestions until a natural break point is detected. Conversely, if the agent infers that a user is in transit or preparing for an important presentation, it might prioritize delivering relevant information or offering timely reminders.

Determining
when and *how* to intervene is a complex design problem. It requires not only robust predictive capabilities but also a nuanced understanding of human workflow and attention spans. Consider the example of an email client that, upon detecting an upcoming meeting, automatically categorizes related emails and

highlights action items or relevant documents within them. This preemptive organization saves the user significant time and mental effort that would otherwise be spent searching and sifting through their inbox. The agent has inferred the user's immediate need – preparation for the meeting – and acted accordingly. The effectiveness of such an intervention hinges on its accuracy and its seamless integration into the user's existing workflow. If the agent misinterprets the situation, or if its actions create additional clutter or confusion, the proactive assistance backfires.

User control is another paramount consideration. While the aim is to automate and anticipate, users must always feel empowered to override or direct the agent's actions. A proactive agent that operates like a black box, making decisions without explanation or recourse, will erode trust and lead to user frustration. Therefore, effective proactive assistance systems must incorporate clear communication and readily accessible controls. When an agent proposes an action, it should clearly articulate
why it's suggesting this action and provide simple mechanisms for the user to accept, reject, or modify the proposed intervention. For instance, a smart home system might detect a user arriving home and suggest turning on specific lights or adjusting the thermostat. The interface should clearly indicate what actions are being proposed and offer immediate options to confirm or dismiss them. This ensures that the agent acts as a helpful co-pilot, not an autonomous overlord.

The design of these communication interfaces is critical.

Instead of overwhelming the user with constant alerts, proactive agents often employ subtle cues or consolidate multiple potential actions into a single, coherent suggestion. A calendar assistant, for example, might not pop up a notification for every single scheduling conflict it detects. Instead, it might consolidate these into a digest presented at a more opportune moment, such as the beginning of the user's workday or a scheduled "planning" block. The communication should be informative, concise, and actionable, reflecting the agent's learned understanding of the user's preferences and priorities. This might involve using natural language to explain the rationale behind an action, such as, "I've noticed you have back-to-back meetings tomorrow afternoon, so I've blocked out 15 minutes for a break before your last appointment. Would you like to keep this break?"

Furthermore, proactive agents can significantly enhance productivity by automating routine and repetitive tasks. Think of a project management tool that, based on task dependencies and team member availability, automatically assigns follow-up actions to individuals as their preceding tasks are completed. This eliminates the need for a manager to constantly track progress and manually reassign work, allowing them to focus on higher-level strategic planning. The agent has learned the project's workflow and is executing it autonomously, freeing up human cognitive resources. This automation extends to areas like data entry, report generation, and even customer support responses. By identifying patterns in user behavior and operational processes, AI agents can take over these predictable tasks, ensuring consistency and reducing the likelihood of human error.

The intelligence behind these agents is often built upon machine learning models that continuously learn and adapt. As the agent interacts with the user and observes their responses to its interventions, it refines its predictive capabilities and its understanding of appropriate action thresholds. This learning process is crucial for long-term effectiveness. An agent that starts out with a somewhat generic approach to proactive assistance will, over time, become highly personalized, aligning its actions with the unique rhythm and preferences of the individual user. For example, a travel assistant might initially suggest booking flights a few weeks in advance. However, after observing that a particular user consistently books flights much closer to the departure date, perhaps due to flexible work arrangements or a preference for last-minute deals, the agent will adjust its timing and suggestions accordingly. This adaptive learning is what transforms a static tool into a dynamic, evolving partner.

The concept of "anticipatory computing" underpins many of these proactive systems. This involves predicting what a user will want or need before they explicitly ask for it, often by analyzing contextual data such as location, time of day, calendar events, communication history, and even physiological cues if available. For instance, a fitness app might proactively suggest a workout based on the user's historical activity patterns, the weather forecast, and their reported energy levels. If the user has a meeting scheduled that requires them to be somewhere specific, the system might proactively calculate travel time, factoring in current traffic conditions, and provide an alert to leave at the optimal moment. This predictive capability is not about mind-reading, but about sophisticated pattern recognition and intelligent inference from a rich stream of data.

Consider the application in smart environments. An AI agent managing a smart home can learn the occupancy patterns of different rooms and the preferred environmental settings for various times of day or activities. When it detects a user entering a room, it can proactively adjust the lighting, temperature, and even ambient music to match their learned preferences. If the agent knows the user typically retires to the bedroom around 10 PM, it can automatically dim the lights in other parts of the house and initiate a pre-set "wind-down" routine. This level of proactive environmental control creates a more comfortable, efficient, and personalized living or working space, all managed seamlessly in the background. The key is that these adjustments are not arbitrary; they are based on a learned model of the user's behavior and preferences, making the experience feel natural and intuitive.

The ethical implications of proactive AI agents are also significant and require careful consideration during the design and deployment phases. One of the primary concerns is the potential for bias embedded within the data used to train these agents. If the historical data reflects societal biases, the agent may learn to make discriminatory decisions. For example, a hiring assistant that proactively suggests candidates for job openings might, if trained on biased historical hiring data, disproportionately recommend candidates from certain demographic groups over others, perpetuating inequality. Rigorous auditing of training data and the development of fairness-aware algorithms are essential to mitigate these risks.

Another ethical challenge revolves around transparency and accountability. When an AI agent acts proactively, it's crucial for users to understand *why* the action was taken. A lack of transparency can lead to distrust and a feeling of being controlled by an inscrutable system. Designers must strive to make the reasoning behind proactive interventions as clear as possible, using plain language explanations and providing audit trails where appropriate. When something goes wrong, it must be clear who or what is responsible, and there should be mechanisms for users to report issues and seek redress.

The potential for over-reliance on proactive systems also raises concerns. If users become accustomed to AI agents handling all the details of scheduling, task management, and decision-making, there's a risk of their own skills in these areas diminishing. This could lead to a reduced capacity for independent problem-solving and a greater vulnerability if the AI system fails or becomes unavailable. Designers should aim to create systems that augment human capabilities rather than replace them entirely, fostering a collaborative relationship where the AI assists and empowers, rather than dictates.

Privacy is another critical ethical consideration. Proactive agents often require access to a wide range of personal data to function effectively. This includes calendar information, communication logs, location data, and potentially even biometric information. Ensuring that this data is collected with informed consent, stored securely, and used only for the intended purpose is paramount. Users must have clear visibility into what data is being collected and how it influences the agent's behavior, along with robust controls to manage their privacy settings and data permissions. The principle of data minimization – collecting only what is strictly necessary – should guide the development of these systems.

The "nudge" effect of proactive assistance also warrants ethical scrutiny. While a gentle nudge can be helpful, the line between helpful suggestion and manipulation can be thin. Proactive agents could be designed to subtly steer users towards certain choices that benefit the provider of the AI service rather than the user. For instance, a travel agent might proactively suggest booking a hotel affiliated with its parent company. Ethical design requires prioritizing user autonomy and well-being, ensuring that proactive suggestions are genuinely in the user's best interest and do not exploit psychological vulnerabilities.

In summary, intelligent agents and proactive assistance represent a significant leap forward in human-computer interaction. By anticipating user needs and taking action on their behalf, these systems have the potential to dramatically enhance productivity, convenience, and personalization. However, realizing this potential requires careful attention to the nuances of user experience, ensuring that interventions are timely, relevant, and respectful of user control. The design of clear communication interfaces, the continuous adaptation of AI models through machine learning, and the thoughtful integration into user workflows are all essential components. Crucially, the development and deployment of these powerful systems must be guided by a strong ethical framework, addressing potential biases, ensuring transparency and accountability, safeguarding privacy, and empowering users to maintain autonomy in an increasingly automated world. The future of intelligent agents lies not just in their ability to predict, but in their capacity to do so responsibly and in partnership with humans.

The integration of Artificial Intelligence into data visualization and exploration represents a paradigm shift in how we interact with and derive meaning from information. Traditionally, data exploration has been a manual, often laborious process. Analysts and researchers would meticulously select datasets, choose appropriate chart types, and iteratively refine their queries to uncover patterns, trends, and outliers. This process, while powerful, demands significant expertise, time, and computational resources. AI, however, promises to democratize this process, making sophisticated data analysis accessible to a broader audience and accelerating the pace of discovery for experts. At its core, AI's contribution lies in its ability to automate complex tasks, learn from data, and even generate insights that might elude human observation.

One of the most transformative applications of AI in this domain is the automated generation of insights. Instead of a user painstakingly crafting queries and visualizations to find specific pieces of information, AI algorithms can sift through vast datasets and proactively identify statistically significant trends, correlations, and anomalies. These insights can range from simple observations, such as a sudden spike in sales for a particular product, to more complex relationships between multiple variables that might not be immediately obvious. For example, in a retail setting, an AI might analyze customer purchase history, demographic data, and even external factors like weather patterns or local events to identify unexpected drivers of demand. It could flag that sales of a specific type of beverage increase significantly on days with high humidity, even if that correlation wasn't a hypothesis being tested. The challenge for designers is not just in the AI's ability to find these insights, but in how to present them to the user in a clear, concise, and actionable manner. Simply presenting a list of statistical findings can be overwhelming. Effective AI-powered visualization tools will curate these insights, prioritizing those that are most relevant and impactful, and providing context or explanations for *why* the AI has flagged them.

Furthermore, AI excels at anomaly detection, a critical component of data exploration in fields ranging from cybersecurity to financial fraud detection and manufacturing quality control. Anomalies, by definition, are data points or patterns that deviate significantly from the norm. Identifying these deviations is crucial for understanding system health, detecting malicious activity, or pinpointing the root cause of problems. AI models, trained on historical data, can establish a baseline of normal behavior and then flag any occurrences that fall outside of this established range. Consider a network security scenario: an AI could monitor network traffic for unusual patterns, such as a sudden surge in data transfer from an unexpected source or a series of failed login attempts from multiple locations. When such anomalies are detected, the AI can trigger alerts, providing immediate notification to security personnel. In manufacturing, AI can analyze sensor data from production lines to identify equipment malfunctions or defects in real-time, preventing the production of large batches of faulty goods. The design of these anomaly detection systems must consider the potential for false positives and false negatives. AI should not only flag anomalies but also provide the user with tools to investigate these flags, such as showing the specific data points contributing to the anomaly and offering historical context.

The selection of appropriate visualizations is another area where AI can significantly enhance the data exploration process. Choosing the right chart type – a bar chart, scatter plot, heatmap, or network graph, for instance – is crucial for effectively communicating data patterns. Different chart types are suited for different kinds of data and different analytical goals. An AI can analyze the dataset's structure, the relationships between variables, and even the user's stated objective to suggest the most informative visualizations. For instance, if a user is exploring the relationship between two continuous variables, the AI might recommend a scatter plot. If they are examining categorical data and its distribution, a bar chart or a pie chart might be suggested. More sophisticated AI can even suggest composite visualizations or dashboards that combine multiple chart types to provide a comprehensive view of complex data. This capability is particularly valuable for users who may not have extensive statistical or visualization expertise. The AI acts as an intelligent guide, steering users towards effective visual representations that facilitate understanding. The design challenge here involves presenting these suggestions in a non-intrusive way and allowing users to easily accept, reject, or further refine the AI's recommendations.

A truly groundbreaking application of AI in data exploration is natural language querying (NLQ). This pattern allows users to interact with data using everyday language, much like asking a question to another person. Instead of learning complex query languages like SQL or building visualizations manually, users can simply type or speak their questions. For example, a marketing manager might ask, "What were our top three selling products in the Northeast region last quarter?" or "Show me the trend of customer acquisition cost over the past year, broken down by marketing channel." The AI engine then translates this natural language query into a structured query that can be executed against the database, and the results are presented, often as a visualization, to the user. This dramatically lowers the barrier to entry for data exploration, empowering business users, domain experts, and even individuals to access and understand data without needing to be data scientists. The effectiveness of NLQ hinges on the AI's ability to accurately understand the intent and context of the user's query, handle ambiguity, and deal with variations in phrasing. Designers must ensure that the NLQ interface provides feedback to the user, confirming their understanding of the query and allowing for clarification if necessary, thereby building trust and improving accuracy.

The presentation and interaction with AI-generated insights are critical design considerations. When an AI identifies a significant trend or anomaly, it's not enough to simply present the finding. Users need to understand the context, the underlying data, and how to act upon the information. This requires interactive elements that allow users to "drill down" into the data that supports the AI's conclusion. For instance, if an AI flags a surge in website traffic, the user should be able to click on that point in the visualization to see the specific referrers, the demographics of the visitors, or the content that was most popular during that period. This ability to explore the provenance of an AI-generated insight is crucial for building trust and enabling deeper analysis. The interface should guide users through this exploration, perhaps by highlighting related data points or suggesting further lines of inquiry based on the user's interactions.

Applications of AI in data visualization and exploration span numerous fields. In business intelligence, AI-powered tools can help identify market trends, understand customer behavior, optimize supply chains, and forecast sales with greater accuracy. For example, a retail business might use AI to analyze customer sentiment from social media and reviews, correlating it with sales data to understand which product features or marketing campaigns are resonating most effectively. This allows for more agile and data-driven decision-making.

In scientific research, AI can accelerate the pace of discovery by analyzing massive datasets from experiments, simulations, or observational studies. For instance, in genomics, AI can help identify genetic markers associated with diseases by analyzing complex patterns across millions of data points. In climate science, AI can analyze satellite imagery and sensor data to identify subtle environmental changes or predict extreme weather events. The ability of AI to detect patterns that might be too complex or subtle for human researchers to spot can lead to groundbreaking insights and new avenues of research.

Personal analytics is another growing area. Fitness trackers, smart home devices, and personal finance apps are increasingly using AI to help individuals understand their own data. An AI might analyze sleep patterns and activity levels to provide personalized health recommendations, or it could track spending habits to suggest budget adjustments and savings opportunities. This empowers individuals to make more informed decisions about their well-being and finances. The ethical considerations around privacy and data security are paramount in these personal applications, as they often involve highly sensitive information.

The design of AI-powered data visualization tools must carefully consider the balance between automation and user control. While AI can automate many steps in the exploration process, users must retain the ability to steer the analysis, question the AI's findings, and apply their own domain expertise. This could involve features that allow users to provide feedback on the relevance of suggested insights, manually adjust parameters for anomaly detection, or even define their own hypotheses for the AI to investigate. A common interaction pattern involves the AI presenting a set of potential insights or visualizations, and the user then selecting the most promising ones for further exploration. This collaborative approach, where AI augments human intelligence rather than replacing it, is key to unlocking the full potential of these tools.

Furthermore, the interpretability of AI models is a significant concern in data visualization and exploration, especially in regulated industries like finance and healthcare. If an AI flags a transaction as fraudulent or recommends a particular treatment based on patient data, stakeholders need to understand *why* that decision was made. This is where AI techniques that provide explanations or justifications for their outputs become crucial. Visualizations can play a vital role in this, by illustrating the data features or patterns that led to a particular AI conclusion. For example, if an AI identifies a patient as being at high risk for a certain condition, a visualization might highlight the specific medical history points, lab results, or genetic factors that contributed to this risk assessment. This transparency is essential for building trust and ensuring accountability.

The continuous learning and adaptation of AI models also present unique design opportunities and challenges. As users interact with AI-powered visualization tools, their behavior and preferences can be observed. This data can be used to refine the AI's understanding of what constitutes a relevant insight, what types of visualizations are most effective for a given user, or how to best interpret natural language queries. This leads to increasingly personalized and effective data exploration experiences over time.

However, this continuous learning also raises questions about data privacy and the potential for the AI to develop unintended biases based on user interactions. Designers must implement mechanisms for users to manage their data, understand how it's being used for learning, and potentially reset or guide the AI's learning process.

In summary, AI is revolutionizing data visualization and exploration by automating insight generation, enhancing anomaly detection, suggesting appropriate visualizations, and enabling natural language querying. These advancements promise to make data analysis more accessible, efficient, and insightful across various domains. The key design challenge lies in creating interfaces that effectively present AI-generated information, empower users to interact with and investigate these findings, and maintain a crucial balance between automation and user control. By focusing on transparency, interpretability, and collaborative interaction, AI-powered tools can truly democratize data and unlock its full potential for discovery and decision-making. The future of data exploration is one where intelligent systems work in partnership with humans, transforming raw data into actionable knowledge with unprecedented speed and clarity.

The culmination of our exploration into essential AI interaction patterns lies in the crucial domain of designing for effective human-AI collaboration. This isn't merely about integrating AI into existing workflows; it's about fundamentally rethinking how humans and intelligent systems can work together, leveraging their distinct yet complementary strengths to achieve outcomes that neither could accomplish alone. The goal is to move beyond a human-versus-AI or AI-for-human paradigm to a true partnership, where the artificial intelligence acts as an intelligent assistant, augmenting human capabilities and freeing up cognitive resources for higher-order tasks. This necessitates a deliberate and thoughtful approach to defining roles, managing the flow of information and control, and ensuring transparent and efficient communication channels between the human and AI components.

At the heart of designing for collaboration is the clear delineation of responsibilities. Understanding what each party – the human and the AI – excels at is paramount. Humans possess invaluable qualities like common sense reasoning, deep contextual understanding derived from lived experience, empathy, ethical judgment, creativity, and the ability to handle novel, ambiguous, or emotionally charged situations. AI, on the other hand, thrives in areas such as processing vast amounts of data at high speed, identifying complex patterns, performing repetitive calculations with unerring accuracy, and recalling information exhaustively. A collaborative design should strategically assign tasks to the agent best suited for them. For instance, in a creative writing context, an AI can be tasked with generating initial drafts, suggesting plot points, or providing a wealth of background research. However, the human writer remains essential for injecting unique voice, nuanced emotional depth, ethical considerations, and the final editorial polish that transforms raw text into compelling prose. The AI acts as a powerful brainstorming partner and content synthesizer, while the human imbues the work with artistry and meaning.

Similarly, consider the realm of medical diagnostics. An AI can meticulously analyze an extensive patient history, cross-reference it with a vast library of medical literature and imaging data, and identify potential diagnoses or flag subtle anomalies that a human clinician might overlook due to fatigue or the sheer volume of information. This AI-driven analysis can present the physician with a ranked list of probable conditions, supported by the evidence it has processed. However, the human doctor's role is indispensable. They bring their clinical judgment, their understanding of the patient's individual circumstances and social determinants of health, their ability to conduct empathetic patient interviews, and their ethical responsibility to make the final treatment decision. The AI provides a sophisticated decision-support mechanism, augmenting the doctor's diagnostic prowess, but it does not replace the critical human element of care, empathy, and ultimate responsibility.

Managing the handoffs between human and AI is another critical aspect of collaborative design. These transitions, where control or information shifts from one agent to the other, must be as seamless and intuitive as possible. A poorly managed handoff can lead to confusion, errors, and frustration, undermining the benefits of collaboration. This requires designing interfaces and workflows that make it evident when a task is being passed to the AI, what information the AI needs to perform its part, and what output to expect. For example, when a financial analyst asks an AI to analyze a complex market trend, the system should clearly indicate that it is processing the request, what data it is using, and when the analysis will be ready. Upon completion, it should present its findings in a comprehensible format, perhaps with interactive elements that allow the analyst to explore the underlying data. The analyst can then take these AI-generated insights and apply their own expertise to formulate strategies, perhaps refining the AI's parameters or asking follow-up questions based on their domain knowledge.

Effective communication is the lifeblood of any successful collaboration, and this holds true for human-AI partnerships. The AI must be able to communicate its actions, its reasoning (to the extent possible), its limitations, and its needs to the human in a clear and understandable manner. Conversely, humans need to be able to provide feedback, corrections, and high-level instructions to the AI. This can manifest in various ways. In natural language querying systems, the AI should confirm its understanding of a user's request before proceeding, and if ambiguity exists, it should prompt for clarification rather than making assumptions. For instance, if a user asks, "Show me the sales data," the AI might respond by asking, "Are you interested in sales by region, product category, or time period?" This simple clarifying question prevents the AI from generating potentially irrelevant information.

Furthermore, the AI's communication should go beyond mere output. When an AI flags an anomaly, for example, it should not just present the anomaly but also offer context and potential explanations. In a cybersecurity scenario, if an AI detects unusual network activity, it should not only alert the security analyst but also provide details about the source of the activity, the type of data being accessed, and the historical patterns that make this activity anomalous. This transparency is crucial for building trust and enabling the human to make informed decisions about how to respond. The human, in turn, might provide feedback to the AI, indicating whether the flagged activity was a false positive, a genuine threat, or something benign but unusual. This feedback loop allows the AI to learn and refine its detection capabilities over time.

The design of collaborative workflows should also consider the varying levels of AI capability and user expertise. Not all AI systems are created equal, and not all users have the same technical proficiency. A robust collaborative system will often offer different modes of interaction. For novice users, a more guided experience might be appropriate, where the AI takes a more proactive role in suggesting actions and interpretations. For expert users, a more open-ended and controllable interface might be preferred, allowing them to fine-tune

AI parameters, set complex constraints, and direct the AI's analysis with greater precision. This adaptability ensures that the collaborative system remains valuable and accessible across a spectrum of users.

Consider the example of an AI-powered legal research assistant. For a junior paralegal, the AI might act as a comprehensive search engine, retrieving relevant case law based on keywords and case summaries, and even suggesting related legal concepts. The paralegal can then review these findings to build a preliminary understanding. For a seasoned litigator, however, the same AI might function as a sophisticated hypothesis-testing tool. The litigator could input specific legal arguments, and the AI would then scour vast legal databases to find supporting or contradicting precedents, analyze judicial trends, and even predict potential outcomes based on historical rulings. The litigator would then use these AI-generated analytical outputs to craft their legal strategy, leveraging the AI's computational power to explore a wider range of legal avenues than would be feasible manually. The handoff here is subtle: the AI performs the intensive search and preliminary analysis, and the litigator takes over for strategic interpretation and argumentation.

The concept of "managed autonomy" is central to designing for human-AI collaboration. This involves granting the AI a degree of autonomy to perform tasks or make decisions within predefined boundaries, while ensuring that the human remains in control and can intervene or override the AI's actions at any point. This is particularly important in safety-critical systems or situations where ethical considerations are paramount. For instance, in an autonomous driving system, the AI is granted autonomy to navigate and control the vehicle, but the human driver always has the ability to take manual control, and the system is designed to alert the driver and seek their input when uncertain situations arise. Similarly, in a medical setting, an AI might suggest a dosage adjustment for a medication, but the final decision to administer that dosage rests with the human physician, who can override the AI if their clinical judgment dictates otherwise.

The interface design plays a pivotal role in facilitating this managed autonomy. Visual cues should clearly indicate when the AI is operating autonomously, what it is currently doing, and what its confidence level is in its actions. Controls for overriding the AI or requesting human intervention should be readily accessible and intuitive. Feedback mechanisms are also crucial. The AI should be able to communicate its rationale for a particular action, especially if it deviates from expected behavior, allowing the human to understand the situation and make an informed decision about intervention. For example, if an AI-controlled robot arm is performing a delicate assembly task and encounters an unexpected obstruction, it should pause, alert the human operator with a visual or auditory cue, and perhaps display an image or video of the obstruction, allowing the operator to remotely guide the robot around the obstacle.

Furthermore, designing for collaboration requires anticipating potential points of failure and designing in resilience. What happens if the AI provides incorrect information or makes a faulty decision? Robust collaborative systems will have mechanisms for error detection and recovery. This could involve having the AI cross-validate its findings with different data sources or algorithms, or designing the workflow so that critical decisions are always reviewed by a human, even if the AI provides a recommendation. The ability for humans to easily report errors or provide corrective feedback to the AI is also vital for continuous improvement. This feedback can be used to retrain the AI models, update their knowledge bases, and improve their performance over time, thereby strengthening the collaborative partnership.

Ultimately, designing for human-AI collaboration is about creating systems that feel like extensions of human capabilities, rather than alien intrusions. It's about building trust through transparency and reliability, empowering users by amplifying their skills, and fostering a symbiotic relationship where the unique strengths of both humans and AI are harnessed for mutual benefit. This requires a deep understanding of human cognition, social dynamics, and the specific context in which the AI will operate. The interaction patterns discussed throughout this chapter—from clear communication and manageable autonomy to effective handoffs and robust feedback loops—all converge on this central theme: designing systems that enable humans and AI to work together, intelligently and effectively, to achieve more than the sum of their parts. The future of intelligent systems is not one of automation replacing humans, but of augmentation, where AI partners with us to unlock new levels of productivity, creativity, and understanding.

Chapter 9: Evaluating AI Systems: Beyond Traditional Metrics

The evolution of human-computer interaction has been marked by increasingly sophisticated evaluation methodologies. For decades, usability testing has been the cornerstone, providing invaluable insights into how users interact with software and hardware. Methods like think-aloud protocols, heuristic evaluations, and standardized task-based testing have been instrumental in identifying design flaws, improving user efficiency, and enhancing overall user satisfaction. These techniques are predicated on a fundamental assumption: that the system under evaluation is largely static and behaves predictably. Users are presented with a defined set of tasks, and their performance is measured against objective metrics such as time on task, error rates, and subjective satisfaction scores. While these traditional approaches have served the field exceptionally well, their efficacy begins to wane when confronted with the unique characteristics of artificial intelligence (AI) systems. The very nature of AI—its dynamism, its capacity to learn and adapt, and its inherent probabilistic and sometimes opaque decision-making processes—introduces complexities that traditional usability testing is ill-equipped to fully address.

One of the primary limitations of traditional usability testing in the context of AI stems from the AI's dynamic and evolving nature. Unlike conventional software, which typically operates with a fixed set of rules and functionalities, AI systems, particularly those employing machine learning, can change their behavior over time.

As an AI system interacts with more data and users, its underlying models are often updated, leading to alterations in its performance, accuracy, and even the way it presents information or takes actions. A usability test conducted on an AI system today might yield results that are no longer valid tomorrow, after the system has undergone further training or adaptation. This inherent mutability means that a static snapshot of user experience, which is what traditional testing often captures, can quickly become outdated. For instance, consider an AI-powered writing assistant. In an initial usability test, users might find its suggestions helpful but occasionally off-topic. However, after the AI has processed more user edits and feedback, its suggestions might become far more relevant and contextually appropriate. A test performed before this learning has occurred would not accurately reflect the system's improved usability. Consequently, the metrics gathered—task completion times, error rates in following suggestions, or satisfaction with the quality of suggestions—might not represent the system's current or future state, rendering the findings less actionable for long-term design and improvement. This presents a significant challenge for establishing reliable benchmarks and for longitudinal studies that track usability improvements over the system's lifecycle.

Furthermore, the learning capability of AI systems can introduce a confounding factor into traditional usability testing. Users are not merely interacting with a fixed interface; they are interacting with a system that is simultaneously learning from their interactions. This learning process can influence the difficulty of tasks, the nature of errors encountered, and the user's perceived efficiency. For example, an AI tutor designed to help students learn a new skill might initially present information in a very structured way. As the student progresses, the AI might adapt its teaching style, offering more complex problems or less guidance. If a usability test focuses solely on the initial stages of this interaction, it might fail to capture how the AI's adaptive capabilities enhance or hinder learning over time. Traditional metrics like "time to complete a module" might be misleading; a longer time initially might indicate a steeper learning curve imposed by the AI, or it could reflect the AI's effective pedagogical approach that necessitates more exploration. Similarly, error rates might decrease not because the user has become more proficient with a static interface, but because the AI has learned to present information in a way that is easier for that particular user to understand. This interplay between user learning and AI learning complicates the interpretation of standard usability metrics. It becomes difficult to disentangle whether a change in performance is due to the user's increasing familiarity with the interface, the AI's adaptation to the user, or a combination of both.

The probabilistic and often opaque nature of AI decision-making, often referred to as the "black box" problem, poses another significant hurdle for traditional usability testing. Many AI systems, especially those based on deep learning, do not operate on deterministic rules that can be easily traced or

explained. When an AI makes a prediction, recommendation, or decision, it is the result of complex computations within its models, the inner workings of which may be inscrutable even to its developers. Traditional usability testing often relies on understanding
why a user made an error or struggled with a task, typically by observing their actions and listening to their explanations. However, when the system itself is behaving unpredictably due to its internal probabilistic nature, it becomes challenging to attribute user difficulties solely to the interface design or user error. For instance, if an AI-powered medical diagnostic tool suggests a particular diagnosis, and the user (a clinician) disagrees or finds the suggestion unconvincing, the "why" behind the AI's suggestion might be impossible to fully articulate. Traditional testing might record this as a user override or a failure of the AI to convince the user, but it cannot easily probe the AI's reasoning to understand the discrepancy. This lack of transparency can erode user trust, a critical factor in the adoption and effective use of AI systems. A user who cannot understand why an AI is suggesting a certain course of action may be hesitant to follow it, even if it is statistically correct. Traditional usability testing, which often focuses on observable behaviors, struggles to adequately measure or address the impact of this opacity on user trust and confidence.

Moreover, traditional usability testing often assumes a single, correct way to perform a task, leading to the expectation of consistent performance across users and sessions. AI systems, however, are designed to handle a far wider range of inputs and to provide diverse outputs, often with inherent variability. For example, an AI image generation tool might produce a visually stunning image based on a text prompt, but the exact rendering can differ with each generation, even with the same prompt. If a usability test aims to measure the "time to generate a satisfactory image," the definition of "satisfactory" becomes highly subjective and can fluctuate based on the AI's output on any given attempt. Furthermore, the goal of the user might not be to achieve a single perfect outcome but to explore possibilities and iterate. Traditional metrics might penalize this exploration by viewing multiple attempts or variations as inefficiencies, when in reality, they are integral to the user's creative process when working with generative AI. The focus on task completion and efficiency, which are core to traditional usability, may overlook the value of exploration, serendipity, and divergent outcomes that are often central to interacting with AI.

The scope of what constitutes "usability" also needs re-evaluation in the context of AI. Traditional usability often focuses on ease of use, learnability, efficiency, and satisfaction with the direct manipulation of an interface. With AI, however, aspects like the perceived intelligence of the system, its trustworthiness, its ability to explain its reasoning (explainability), and its fairness become equally, if not more, important. An AI system might be technically easy to operate, but if it is perceived as unfair, biased, or untrustworthy, its overall usability will be severely compromised. Traditional methods like user surveys (e.g., SUS, PSSUQ) are useful for capturing subjective satisfaction, but they may not have the granularity to probe these deeper, AI-specific concerns effectively. For instance, a user might rate an AI customer service chatbot as "easy to use" because it understands their queries, but if they suspect the chatbot is prioritizing upselling over genuine problem-solving, their overall experience will be negative, and their trust will be diminished. Capturing this nuanced negative sentiment through traditional satisfaction metrics alone can be difficult.

The reliance on representative user groups in traditional usability testing also faces challenges with AI. AI systems can have a wide range of applications, and the expertise of users can vary dramatically. For example, a medical AI system might be used by highly trained physicians, nurses, and potentially even patients. Each group will have different needs, expectations, and levels of technical understanding. Traditional testing might struggle to recruit and effectively test with such a diverse set of users simultaneously, or to design tasks that are equally relevant and challenging for all. Furthermore, the performance of an AI system can be highly dependent on the specific data it has been trained on and the context in which it is deployed. A system that performs well with data from one hospital or demographic might perform poorly with data from another. Traditional usability tests, often conducted in controlled laboratory settings, may not adequately capture these domain-specific performance variations or the impact of data drift over time, which can significantly affect the perceived usability and utility of the AI.

In essence, while traditional usability testing provides a valuable foundation, its limitations become apparent when confronted with the adaptive, probabilistic, and often opaque nature of AI systems. The assumption of static behavior, the difficulty in disentangling user learning from system adaptation, the challenges in explaining AI's "black box" decisions, and the broader scope of what constitutes "usability" for intelligent systems all point to the need for a paradigm shift in how we evaluate AI. Merely applying existing methods might lead to incomplete or misleading conclusions, hindering the development of truly effective, trustworthy, and user-centered AI. This necessitates the exploration and adoption of evaluation techniques that are specifically designed to address the unique characteristics of AI, moving beyond simple efficiency and satisfaction metrics to encompass trust, transparency, fairness, and the dynamic interplay between human and intelligent machine.

The challenge of evaluating usability in adaptive AI systems necessitates a departure from conventional metrics that assume a static user interface and predictable user behavior. Instead, we must embrace methodologies that can capture the fluid, dynamic, and often personalized interaction that characterizes these intelligent systems. This involves considering how the system's evolution impacts the user experience and developing new ways to quantify that impact.

One critical aspect is the concept of **learnability** within an adaptive context. Traditional learnability measures often focus on how quickly a user can become proficient with a system's initial features and functionalities. However, in an adaptive system, learnability takes on a dual meaning. It not only refers to the user's ability to learn the system's initial operations but also their capacity to adapt to and effectively utilize the system as it changes and evolves. This means evaluating not just the initial onboarding but also the user's ongoing ability to understand and leverage new or modified behaviors that the AI introduces. For instance, a user might initially learn to operate an AI-powered scheduling assistant by manually inputting meeting details. As the AI learns the user's preferences and common collaborators, it might begin to proactively suggest meeting times or even initiate calendar invites. A measure of learnability in this adaptive scenario would assess how quickly and effectively the user can transition from manual input to leveraging these automated suggestions, or how easily they can correct the AI's erroneous assumptions when they arise. This requires observing how users respond to the AI's evolving capabilities, noting the time and effort it takes for them to adjust their interaction patterns, and measuring their success in achieving their goals with the increasingly sophisticated system.

Beyond learnability, the very **adaptability** of the AI itself needs to be evaluated from a user-centric perspective. While technical metrics can quantify how well an AI adapts to new data or user input, the crucial question for usability is how this adaptation is perceived and experienced by the user. Does the AI's adaptation make the system easier or harder to use? Does it align with the user's implicit expectations or disrupt their workflow? Evaluating user-perceived adaptability involves understanding the rate and nature of changes the user encounters and their comfort level with these alterations. This could involve tracking how frequently users encounter unexpected system behaviors and how readily they can re-establish their understanding of how to interact with the system. For example, consider an AI-driven personal finance manager that adjusts its budgeting recommendations based on a user's spending habits. If the AI suddenly shifts its categorization of certain expenses or alters its savings targets drastically without clear explanation, it might lead to user frustration and a perceived decrease in usability, even if the AI's underlying models are technically improving. Usability evaluation in such cases would need to assess not just the accuracy of the new recommendations but also the user's understanding of the changes and their ability to provide corrective feedback that the AI can effectively incorporate. This might involve surveys asking users how surprised they were by system changes, how well they understood the reasons behind them, and whether they felt in control of the system's evolving behavior.

User satisfaction in adaptive systems is also a far more nuanced construct. It's not simply about whether the user likes the system at a single point in time, but whether their satisfaction is sustained or even grows as the system learns and adapts. A key element here is the concept of **trust**. As an AI system becomes more personalized and takes on more complex tasks, users need to trust its recommendations and actions. This trust is built and eroded by the AI's performance and its transparency. If an adaptive system consistently provides helpful and accurate responses, user trust and satisfaction are likely to increase. Conversely, if it makes errors, exhibits biases, or acts in ways that are difficult to understand, trust will diminish, leading to dissatisfaction. Measuring this evolving trust can be achieved through longitudinal studies, where user perceptions are tracked over extended periods. This might involve regular check-ins with users, asking them to rate their confidence in the AI's outputs, their willingness to rely on its suggestions, and their overall comfort level with its behavior. For instance, an AI-powered content recommendation engine that learns a user's viewing habits might initially offer accurate suggestions. As it adapts, however, it might start to over-specialize, creating a "filter bubble" or, conversely, introduce irrelevant content as it attempts to broaden its understanding. The user's satisfaction will hinge on whether these adaptations enhance or detract from their overall content discovery experience, and their trust in the system's ability to genuinely serve their interests.

To quantitatively assess these evolving aspects, new metrics and methodologies are required. Beyond traditional task completion times and error rates, we can consider metrics that capture the

rate of adaptation itself from a user's perspective. This could involve measuring the time it takes for a user to achieve a certain level of proficiency with an updated version of the system, or the number of corrective actions a user needs to take after an AI adaptation. For example, in an adaptive medical diagnostic support system, we might measure how long it takes a physician to adjust their diagnostic workflow after the AI has been updated with new training data that alters its diagnostic probabilities. This would involve tracking the physician's decision-making process and comparing it to previous performance benchmarks. Another metric could be the **stability of user performance** in the face of AI changes. Does the user's efficiency and accuracy remain consistent, or does it fluctuate significantly as the AI adapts? A system that allows for stable user performance despite its own internal evolution would be considered more usable.

Furthermore, subjective measures need to be tailored to capture the unique challenges of adaptive AI. Standard questionnaires like the System Usability Scale (SUS) or Post-Study Usability Questionnaire (PSUQ) might provide a general sense of satisfaction, but they often lack the granularity to probe specific issues related to AI adaptation. We might need to develop new survey items that directly ask users about their perceptions of the AI's learning process, their comfort with its evolving behavior, and their trust in its decisions. Questions could include: "How often does the system behave in a way that surprises you?", "How well do you understand why the system has changed its behavior?", and "How confident are you in the system's recommendations after its recent updates?". The **perceived intelligence** of the system is another crucial, albeit difficult, dimension to measure. While not a traditional usability metric, a user's perception of the AI's intelligence directly influences their engagement and satisfaction. This can be assessed through qualitative interviews and by analyzing user-generated narratives about their experiences with the system.

The integration of
explainability (XAI) directly into usability evaluation is paramount. When an AI system adapts, understanding *why* it has changed its behavior is often as important as the change itself for maintaining user trust and facilitating adaptation. Usability evaluations should therefore include measures of how well users understand the AI's reasoning behind its adaptive actions. This can be assessed by asking users to predict the AI's behavior or to explain its actions, and then comparing their responses to the actual AI logic. If the AI's explanations are unclear, incomplete, or inaccurate, it will negatively impact usability. For example, if an AI writing assistant suddenly starts recommending a different tone or vocabulary, and it provides a clear explanation tied to the document's context or the user's stated goals, the user is more likely to accept and learn from the suggestion. Without such an explanation, the user might dismiss the suggestion as arbitrary or even intrusive. Evaluating the effectiveness of these explanations becomes a direct component of usability for adaptive AI.

Moreover, the evaluation of adaptive systems needs to acknowledge the
user's agency and control. While the AI is adapting, users should ideally retain a sense of control over the system and its behavior. Usability metrics can be designed to assess the perceived level of control users have. This might involve measuring how easy it is for users to override the AI's decisions, to provide feedback that influences its adaptation, or to revert to previous system states if necessary. A system that empowers users to guide or constrain the AI's adaptation, rather than being passively subjected to it, will generally be perceived as more usable and trustworthy. For instance, an adaptive navigation system that learns a user's preferred routes and driving style should offer options for the user to explicitly set preferences, such as "avoid highways" or "prefer scenic routes," and to easily correct the AI if it suggests an undesirable path. The ease with which users can exercise this control is a critical usability factor.

Finally, the
longitudinal nature of user interaction with adaptive systems demands longitudinal evaluation approaches. Static usability tests performed at a single point in time will inevitably fail to capture the full user experience. Instead, we need methods that track how usability evolves alongside the AI's adaptation. This could involve repeated usability testing sessions over weeks or months, or the deployment of in-situ monitoring tools that collect data on user behavior and system performance in real-world contexts. Analyzing trends in

user satisfaction, task efficiency, error rates, and trust over time will provide a much richer and more accurate picture of the adaptive system's usability. This ongoing evaluation allows designers to identify points where the AI's adaptation might be causing friction, to refine the adaptation algorithms based on real user feedback, and to ensure that the system remains usable and valuable throughout its lifecycle.

In summary, measuring usability in adaptive systems is a complex but essential endeavor. It requires moving beyond traditional, static metrics to embrace a more dynamic and user-centered approach. By focusing on learnability in an evolving context, user-perceived adaptability, sustained user satisfaction built on trust, and the effective integration of explainability and user control, we can begin to develop robust methodologies for evaluating the usability of AI systems that learn and change over time. This shift is not merely an academic exercise; it is fundamental to building AI that is not only technically advanced but also genuinely helpful, trustworthy, and a pleasure to interact with.

The evaluation of AI systems must extend beyond traditional performance metrics and usability considerations to encompass their fairness and broader ethical impact. While technical accuracy and efficiency are crucial, an AI system's societal contribution hinges on its equitable treatment of all users and its alignment with fundamental human values. This subsection delves into the practical, real-world assessment of these critical dimensions, moving from abstract principles to tangible methodologies for ensuring AI systems are not only functional but also just and responsible.

One of the most significant challenges in deploying AI in real-world scenarios is the potential for **bias drift**. Unlike static software, AI models are often designed to learn and adapt based on incoming data. While this adaptability is a strength for performance enhancement, it can also be a breeding ground for emergent unfairness. As an AI system interacts with a population, the data it receives may not be representative of all demographic groups, or it may reflect existing societal biases. Over time, these imbalances can be amplified, leading the AI to make decisions or predictions that systematically disadvantage certain groups. Therefore, continuous, proactive monitoring is essential. This involves establishing robust data pipelines that not only feed the AI but also track the demographic composition and statistical properties of the data it is learning from. Sophisticated anomaly detection algorithms can be employed to flag significant deviations in data distribution that might indicate a drift towards bias. For instance, a hiring AI that initially demonstrated parity in screening candidates from different ethnic backgrounds might, after months of operation, begin to disproportionately favor candidates from a majority group if its training data predominantly comprises applicants from that group. Evaluating this requires not just periodically retraining the model but continuously auditing the input data and the model's output distributions across defined protected attributes (e.g., race, gender, age, socioeconomic status). This monitoring should be granular, looking for disparities in false positive rates, false negative rates, prediction accuracy, and recommendation diversity for different user segments.

Beyond algorithmic monitoring, understanding how users
perceive fairness is a crucial, often overlooked, aspect of ethical evaluation. What constitutes fairness can be subjective and context-dependent. A system might be technically fair according to predefined metrics, yet users from marginalized communities might still perceive its outcomes as unjust. This highlights the necessity of incorporating user perception studies into the evaluation framework. These studies can take various forms, from qualitative interviews and focus groups to structured surveys designed to elicit user attitudes towards the AI's decisions. For example, in a loan application AI, even if approval rates are statistically equitable, users who are denied loans might feel the process is unfair if they don't understand the criteria used or believe the system is implicitly biased against their demographic profile. User perception studies can uncover these nuances. Researchers might present users with hypothetical scenarios of AI decision-making and ask them to rate the fairness of the outcome, or to explain their reasoning. Longitudinal studies are particularly valuable here, as they can track how user perceptions of fairness evolve over time as they interact more with the system and observe its consistent or inconsistent application of rules. Such studies can reveal whether a system's adaptive

learning, intended to improve outcomes, is inadvertently creating new forms of perceived inequity.

Auditing AI decision-making processes in real-world scenarios is another indispensable practice for evaluating fairness and ethical impact. This goes beyond inspecting the code or the model's parameters; it involves examining the actual outcomes of the AI in operation. For systems that make high-stakes decisions – such as in criminal justice, healthcare, or finance – a rigorous audit trail is paramount. This means meticulously documenting the inputs, the intermediate decision-making steps (where possible through explainable AI techniques), and the final outputs for each instance. Post-hoc analysis of these logs can reveal patterns of bias that might not be apparent through simpler monitoring. For example, an AI used for sentencing recommendations might appear unbiased when looking at overall conviction rates, but an audit could reveal that it assigns harsher recommendations for similar offenses when committed by individuals from certain racial groups. This requires not only technical expertise in data analysis but also domain knowledge to understand what constitutes a fair outcome within that specific context. These audits should not be a one-time event but a regular occurrence, integrated into the operational lifecycle of the AI system. They should be conducted by independent bodies or internal ethics committees with the authority to recommend or mandate changes.

To effectively steer AI development and deployment towards ethical ends, establishing clear **ethical Key Performance Indicators (KPIs)** is crucial. These KPIs translate abstract ethical principles into measurable targets. Unlike traditional performance metrics like accuracy or latency, ethical KPIs focus on equity, accountability, and societal well-being. Examples of ethical KPIs could include: the reduction of disparity in outcomes across protected groups (e.g., ensuring the difference in loan approval rates between two demographic groups does not exceed a predefined threshold), the rate at which the AI's decisions are successfully appealed or overturned, the level of transparency achieved in explaining AI decisions to users, or metrics related to user privacy and data security. For instance, a content moderation AI might have a KPI related to the equitable application of community guidelines, measuring whether content from different political affiliations is removed at similar rates for similar violations. The process of defining these KPIs requires careful consideration and multi-stakeholder input, involving ethicists, legal experts, domain specialists, and representatives of the communities most likely to be affected by the AI.

Regularly assessing an AI's **alignment with societal values and user well-being** is an ongoing imperative. AI systems operate within complex social ecosystems, and their impact extends beyond individual interactions to broader societal implications. This assessment necessitates a holistic view that considers potential harms, benefits, and unintended consequences. For instance, an AI designed to optimize traffic flow might improve commute times but could inadvertently exacerbate pollution in certain neighborhoods or displace residents due to new infrastructure. Evaluating this requires broad societal impact assessments, potentially involving participatory design workshops where diverse stakeholders can voice concerns and contribute to the AI's ethical roadmap. It also involves establishing feedback loops from end-users, support staff, and the public to capture emergent issues. For a healthcare AI, assessing alignment with societal values might mean evaluating its impact on healthcare access for underserved populations, its contribution to clinician burnout, or its adherence to patient autonomy and privacy. This assessment process is inherently iterative; as societal norms evolve and new ethical challenges emerge, the AI's alignment must be re-evaluated and adjusted. This continuous process ensures that the AI system remains a force for positive and equitable change, rather than an unintended source of harm or disparity. The ultimate goal is to ensure that AI systems are not merely efficient tools but are developed and deployed in a manner that upholds human dignity, promotes justice, and contributes positively to the collective good.

The dynamic nature of Artificial Intelligence necessitates a paradigm shift in how we approach its evaluation and refinement. While initial performance metrics and ethical audits provide a crucial baseline, the true measure of an AI system's success, particularly in its long-term impact and alignment with human values, lies in its continuous evolution. This evolution is not solely driven by algorithmic advancements or developer intuition, but fundamentally by the lived experiences and expressed needs of its users. User feedback, therefore, is not an ancillary consideration but the very engine that drives iterative AI development towards genuine usefulness and ethical integrity. It transforms the AI from a static product into a responsive partner, capable of adapting to the nuanced realities of its deployment.

The collection of user feedback must be a systematic and integrated process, not an afterthought. This involves designing AI systems with inherent mechanisms for capturing user input throughout their lifecycle. For instance, within a conversational AI, a simple "Was this helpful?" prompt after a response, coupled with an option to provide more detailed feedback, can offer a wealth of information. This feedback can range from identifying factual inaccuracies or irrelevant suggestions to highlighting instances where the AI's tone was inappropriate or its understanding was flawed. Beyond direct questions, implicit feedback can be gleaned from user behavior. If a user repeatedly rephrases a query, ignores an AI's recommendation, or navigates away from a feature after a few interactions, these are all signals indicating that the AI is not meeting their needs effectively. These signals, when aggregated and analyzed, can point to areas where the AI's comprehension, reasoning, or output requires refinement.

One of the most powerful techniques for understanding the qualitative nature of user feedback is

sentiment analysis. This natural language processing (NLP) technique allows us to automatically process and categorize large volumes of user-generated text—such as reviews, support tickets, or social media comments—to gauge the emotional tone behind them. For an AI powering a customer service chatbot, sentiment analysis can quickly flag interactions where users expressed frustration, confusion, or dissatisfaction. Identifying these negative sentiment patterns in real-time allows development teams to prioritize issues that are causing the most user distress. For example, if multiple users express anger after being unable to complete a transaction due to an AI error, this can be immediately escalated, prompting a rapid investigation and fix. Conversely, positive sentiment can highlight areas where the AI is performing exceptionally well, offering insights into successful design patterns or effective communication strategies that can be replicated elsewhere. The granularity of sentiment analysis can even extend to identifying specific aspects of the AI that elicit positive or negative reactions, such as the clarity of its explanations or the speed of its responses.

Beyond sentiment,
detailed issue tracking is paramount for translating user feedback into actionable improvements. This involves establishing robust systems for logging, categorizing, and prioritizing bugs, usability problems, and feature requests reported by users. Each piece of feedback should be treated as a data point that can inform the iterative development cycle. For a machine learning model used in content recommendation, user

feedback might highlight instances where irrelevant or even offensive content was suggested. A well-structured issue tracking system would allow developers to link these user reports to specific instances of the AI's output, examine the input data that led to that output, and implement corrective measures. This could involve retraining the model with more diverse data, adjusting the recommendation algorithm, or introducing new filtering mechanisms. The key is to create a closed-loop system where feedback directly informs the problem-solving process and where the impact of changes made in response to feedback is subsequently monitored. This process ensures that the AI is not just being improved, but is being improved in directions that are demonstrably valuable to its users.

A/B testing plays a critical role in the iterative refinement of AI features and algorithms based on user feedback. Once potential improvements have been identified, A/B testing allows developers to deploy different versions of an AI feature to distinct segments of the user base and quantitatively measure which version performs better according to predefined metrics. For an AI that generates personalized learning paths, developers might hypothesize that a new algorithm for sequencing topics will lead to higher engagement. They could then deploy this new algorithm (version B) to a subset of users while a control group (version A) continues to use the existing algorithm. By tracking metrics such as completion rates, time spent on modules, and user satisfaction scores, the developers can objectively determine if version B is indeed superior. This data-driven approach minimizes reliance on assumptions and ensures that changes are implemented based on empirical evidence of user preference and efficacy. The insights gained from A/B tests can then inform further iterations, creating a continuous cycle of experimentation and optimization. This methodology is particularly valuable when dealing with complex AI behaviors where human intuition might be insufficient to predict the optimal approach.

Establishing
robust feedback loops is essential for maintaining alignment between the AI system and its users' evolving needs and expectations. This means creating multiple, accessible channels for users to provide input and ensuring that this input is regularly reviewed and acted upon by the development team. These channels can include in-app feedback forms, dedicated user forums, regular surveys, and even direct engagement through beta testing programs. For example, an AI-powered writing assistant might offer users the ability to rate the quality of its suggestions or report when it has misunderstood their intent. The feedback from these interactions should be systematically collected, analyzed for recurring themes, and then used to prioritize updates to the AI's underlying models and functionalities. Furthermore, it is crucial to communicate back to users about how their feedback has been incorporated. This not only fosters a sense of community and engagement but also reinforces the value of their contributions, encouraging continued participation. Transparency about the iterative process,

even at a high level, builds trust and helps manage user expectations.

The impact of iterative AI development driven by user feedback extends beyond mere performance enhancement; it is deeply intertwined with **ethical considerations**. When users can easily report instances of bias, unfairness, or discriminatory behavior from an AI, they become active participants in its ethical governance. Consider an AI used for loan applications. If users from certain demographics consistently report feeling that the AI's decisions are opaque or biased, this feedback can trigger an investigation into potential systemic inequities within the model or its training data. Without this direct user input, such issues might remain hidden, perpetuating harm. By actively soliciting and responding to user feedback regarding fairness, organizations can proactively identify and mitigate biases that might have been overlooked during initial development or that have emerged due to drift in real-world data. This user-centric approach to ethical evaluation ensures that the AI's development remains grounded in principles of equity and justice, as perceived and experienced by the diverse individuals it serves.

Furthermore, **longitudinal studies** that track user feedback over extended periods are invaluable for understanding the long-term impact of AI systems and their evolving relationship with users. As users become more familiar with an AI, their expectations and feedback patterns may change. What initially seemed novel or helpful might, over time, become commonplace, or new frustrations might emerge. Longitudinal studies allow researchers to observe these shifts and adapt the AI accordingly. For instance, a personal finance AI that initially impressed users with its budgeting tools might, after a year of use, face criticism for lacking advanced investment advice or sophisticated tax planning features. By monitoring this evolving user sentiment, developers can proactively plan for future iterations, ensuring the AI remains relevant and valuable throughout its lifespan. This ongoing dialogue ensures that the AI doesn't become obsolete or irrelevant as user needs and technological landscapes change.

The integration of user feedback into the AI development lifecycle also necessitates **sophisticated analytical tools**. Beyond simple sentiment analysis, techniques like topic modeling can identify recurring themes and concerns within large volumes of unstructured feedback. This allows developers to pinpoint specific areas of the AI that are consistently causing problems or eliciting praise. For example, topic modeling applied to feedback for a virtual assistant might reveal a recurring theme of "difficulty understanding accents" or "inconsistent performance with complex commands." These insights can then guide targeted improvements to the AI's speech recognition modules or its natural language understanding capabilities. Similarly, **clustering algorithms** can group similar feedback entries together, helping to identify common pain points that might affect a significant portion of the user base. This allows for the prioritization of fixes that will have the broadest positive impact.

The **responsiveness** to user feedback is as critical as its collection. A system that collects feedback but fails to act upon it will eventually lead to user disengagement and frustration. Therefore, organizations deploying AI systems must establish clear processes and allocate resources for analyzing feedback, prioritizing necessary changes, implementing them, and then communicating these updates back to the user community. This commitment to iteration and responsiveness is what builds trust and fosters a positive relationship between users and the AI. It transforms users from passive recipients of technology into active collaborators in its development. This can involve dedicated user research teams, agile development methodologies that incorporate regular feedback cycles, and transparent communication strategies about upcoming features and improvements.

Ultimately, the iterative refinement of AI systems through user feedback is a fundamental practice for ensuring that these powerful technologies are not only technically sound but also deeply aligned with human needs, values, and ethical considerations. It is through this continuous dialogue—listening to, understanding, and acting upon the voices of those who interact with the AI—that we can steer its development towards a future where AI serves humanity equitably and effectively, fostering innovation that is both intelligent and deeply human. This approach recognizes that AI is not a finished product to be delivered, but a continuously evolving entity that thrives on engagement and adaptation, much like any living system.

The true measure of an artificial intelligence system's maturity and trustworthiness extends far beyond its initial performance benchmarks or adherence to ethical guidelines. Once an AI is deployed into the real world, it faces an unpredictable torrent of conditions that can challenge its design and intended functionality. Therefore, a critical aspect of evaluating AI systems, particularly those destined for high-stakes applications, is a rigorous examination of their **robustness and reliability**. This involves moving beyond average-case performance to understand how the AI behaves when faced with deviations from its ideal operating environment.

Robustness, in the context of AI, refers to the system's ability to maintain its performance and safety even when subjected to inputs that differ from the training data, or when operating under unusual or stressful conditions. This is not merely about preventing outright failure; it's about ensuring graceful degradation or, ideally, continued competent operation. Reliability, on the other hand, speaks to the consistency and dependability of the AI's outputs over time and across different situations. An AI can be robust in the face of minor input variations but still be unreliable if its decisions are erratic or prone to unexpected shifts. For systems involved in patient diagnosis, autonomous navigation, or financial forecasting, any lapse in robustness or reliability can have severe consequences, ranging from inconvenience to catastrophic harm.

One of the foundational approaches to assessing robustness is
stress testing. This involves pushing the AI system to its limits by exposing it to an overload of data, extreme input values, or rapidly changing environmental variables. For a machine learning model tasked with image recognition, stress testing might involve feeding it images with extreme lighting conditions, low resolution, or significant occlusion. For a natural language processing system, this could mean processing exceptionally long sentences, highly colloquial language, or a barrage of rapid-fire queries. The goal here is not to find an optimal operating point, but to identify the breaking points and understand the nature of the failure. Does the system crash? Does it produce nonsensical outputs? Or does it simply become less accurate, and by how much? The insights gained from stress testing are invaluable for setting operational boundaries, understanding potential failure modes, and informing the design of fail-safe mechanisms. For instance, if stress testing reveals that an AI image classifier becomes highly unreliable when more than 70% of an object is obscured, developers can implement a confidence threshold that flags such images for human review, rather than allowing a potentially incorrect classification to proceed.

Complementing stress testing is

fault injection. This technique deliberately introduces errors or anomalies into the AI's inputs, internal states, or even its operational environment to observe its reaction. This can range from subtly corrupting individual data points within a dataset to simulating hardware malfunctions or network disruptions. For an AI controlling a robotic arm, fault injection might involve simulating a sensor failure, a loss of communication with a joint actuator, or a sudden increase in friction. The AI's response—whether it halts operations, attempts a recovery maneuver, or enters a safe mode—reveals its fault tolerance. Similarly, for a financial trading AI, fault injection could simulate sudden market volatility, corrupted transaction data, or delays in receiving critical information. Analyzing how the AI handles these simulated faults provides critical data on its resilience and its ability to maintain a safe or predictable state during unexpected events. This proactive identification of vulnerabilities allows for the implementation of more robust error handling, redundancy, and recovery protocols.

Beyond injecting faults,
scenario-based testing offers a more holistic approach to evaluating robustness and reliability by simulating real-world operational contexts. This involves crafting specific, plausible scenarios that the AI is likely to encounter during its deployment. For an autonomous vehicle's AI, these scenarios might include navigating a busy intersection with unpredictable pedestrian behavior, responding to emergency vehicles, or encountering sudden adverse weather conditions. For a medical diagnostic AI, scenarios could involve

processing patient data with missing information, interpreting ambiguous imaging results, or dealing with rare but critical conditions. The key is to design scenarios that are representative of the complexities and uncertainties of the target environment. These scenarios are often developed through domain expertise, user feedback, and historical data analysis. Evaluating the AI's performance across a diverse set of well-defined scenarios allows developers to build confidence that the system can handle a wide range of real-world situations. It also helps to uncover subtle interactions between different components of the AI system or between the AI and its environment that might not be apparent in simpler testing methods.

A particularly insidious threat to AI robustness comes from
adversarial attacks. These are deliberate attempts to mislead or manipulate an AI system by crafting specific inputs that are subtly altered from legitimate data. To a human observer, these altered inputs might appear almost identical to the original, but they can cause an AI to make drastically incorrect classifications or predictions. For example, a few carefully placed pixels added to an image of a stop sign can cause an autonomous vehicle's vision system to misclassify it as a speed limit sign, with potentially fatal consequences. Similarly, minor perturbations in text can cause a sentiment analysis model to flip its interpretation from positive to negative, or a spam filter to let malicious emails through. Evaluating an AI's resilience to adversarial attacks is crucial, especially for security-sensitive applications. This often involves not only

testing against known attack methodologies but also developing new defenses, such as input sanitization, adversarial training (exposing the model to adversarial examples during training to teach it to resist them), and robust detection mechanisms. The arms race between attackers and defenders in this domain highlights the ongoing challenge of securing AI systems against malicious manipulation.

The nature of the data fed into an AI system can also significantly impact its robustness and reliability. **Noisy data**, characterized by errors, inaccuracies, or extraneous information, is a common challenge in real-world applications. This can arise from faulty sensors, human data entry errors, or limitations in data collection processes. An AI system that is not designed to handle noisy data may produce inaccurate or unreliable outputs, or even fail entirely when confronted with it. Techniques for evaluating robustness in the presence of noisy data include:

Data Perturbation Testing: Similar to fault injection but specifically focused on data. This involves systematically adding different types and levels of noise to the training and testing datasets to observe how the AI's performance degrades. This can include Gaussian noise, salt-and-pepper noise for images, or word deletions/substitutions for text.
Outlier Detection and Handling: Evaluating the AI's ability to identify and appropriately handle data points that are significantly different from the rest of the dataset. Does it flag them for review? Does it ignore

them? Or does it let them unduly influence its decisions?

Robust Training Methodologies: Assessing whether the AI has been trained using methods that inherently promote robustness to noise. Techniques like data augmentation (creating artificial variations of training data) or using robust loss functions can improve resilience.

The evaluation of robustness and reliability is not a one-time event but an ongoing process. As AI systems learn and adapt, their behavior can change, and new vulnerabilities may emerge. Therefore, **continuous monitoring and re-evaluation** are essential. This involves setting up systems to track the AI's performance in production, detect anomalies, and gather data on how it is behaving under real-world conditions. User feedback, though qualitative, can also serve as an early warning system for reliability issues. A sudden increase in user complaints about incorrect recommendations or unexpected behavior can signal underlying problems with robustness.

Furthermore, the complexity of modern AI systems, particularly deep neural networks, can make it difficult to understand *why* a system fails. This lack of **explainability** poses a significant challenge for debugging and improving robustness. When an AI makes an incorrect prediction or behaves unexpectedly, understanding the reasoning behind that behavior is crucial for fixing it. This is where techniques in **Explainable AI (XAI)** become intertwined with robustness testing. By employing XAI methods, developers can gain insights into which features or inputs most influenced an AI's decision, thereby helping to pinpoint the source of errors or vulnerabilities. For example, if an image classifier incorrectly identifies a cat as a dog, XAI techniques might reveal that an unusual shadow or a specific background element unduly influenced the decision, allowing developers to address this bias or noise sensitivity.

The deployment context is paramount in defining what constitutes acceptable robustness and reliability. An AI for a video game might tolerate a higher degree of occasional glitching than an AI used in a medical device. Therefore,
risk assessment must inform the rigor and scope of robustness and reliability testing. For systems with high potential impact on human safety or well-being, such as autonomous driving systems or AI in critical infrastructure, the bar for robustness and reliability must be exceedingly high, requiring extensive validation across a vast array of challenging conditions. This might involve simulated environments that mirror real-world complexity with high fidelity, as well as carefully controlled real-world testing.

In summary, evaluating AI system robustness and reliability is a multi-faceted endeavor that moves beyond standard performance metrics. It requires a proactive and comprehensive approach encompassing stress testing, fault injection, scenario-based evaluations, and defense against adversarial attacks. It necessitates understanding how the system behaves with imperfect data and developing mechanisms for continuous monitoring and improvement. By rigorously assessing and enhancing these critical aspects, we can build greater confidence in the dependable operation of AI systems, paving the way for their safe and effective integration into increasingly complex and vital domains of our lives. This commitment to dependability is not merely a technical challenge; it is a fundamental requirement for fostering trust and ensuring that AI technologies serve humanity as intended, without succumbing to the inherent uncertainties of the real world.

Chapter 10: Longitudinal Studies and Long-Term Impact

The imperative to rigorously assess artificial intelligence systems extends well beyond the immediate performance metrics captured in laboratory settings or during initial deployment phases. While short-term evaluations, user studies, and stress tests are crucial for understanding an AI's immediate capabilities, robustness, and reliability, they often provide an incomplete picture of its true impact. The ways in which humans interact with, adapt to, and are influenced by AI technologies unfold gradually, over months and years, subtly reshaping behaviors, cognitive processes, and even societal structures. To truly understand and responsibly govern AI, a fundamental shift in evaluation methodology is required: the adoption of longitudinal studies. These studies are designed to track the evolution of AI systems and their users over extended periods, revealing a richer, more nuanced, and ultimately more critical understanding of their long-term consequences.

The limitations of short-term evaluations become starkly apparent when considering the dynamic nature of human-AI interaction. When users first encounter a new AI system, their behavior is often characterized by a learning curve. They explore its functionalities, test its boundaries, and develop initial strategies for engagement. This period is heavily influenced by novelty, curiosity, and the explicit guidance provided by the system or its designers. Metrics captured during this phase might reflect an initial positive reception or a rapid mastery of basic tasks. However, this snapshot fails to capture what happens *after* the novelty wears off, once the AI becomes a routine part of a user's life or workflow. Do users continue to engage with the system's full capabilities, or do they default to a simplified, perhaps less effective, mode of interaction? Does the AI's presence lead to a gradual erosion of certain human skills or a development of over-reliance that could be detrimental in situations where the AI is unavailable or malfunctions? These are questions that can only be answered by observing users and systems over significant durations.

Consider the example of an AI-powered writing assistant designed to improve grammar, style, and clarity. In a two-week user study, participants might report higher satisfaction with their writing and a perceived improvement in their output, as the AI actively guides them towards more polished prose. The system's immediate metrics would likely reflect a reduction in grammatical errors and potentially faster drafting times. However, a longitudinal study conducted over six months or a year might reveal a different story. Users, consistently relying on the AI's suggestions, might begin to internalize its stylistic preferences without fully understanding the underlying grammatical rules. This can lead to a phenomenon known as "skill atrophy," where the user's own inherent writing abilities diminish because the AI is doing the heavy lifting. They might find themselves unable to produce coherent text without the assistant, or their own critical judgment regarding sentence structure and word choice might weaken. Furthermore, the AI might inadvertently homogenize writing styles, suppressing individual voice and creativity in favor of a universally acceptable, yet bland, output. The short-term gains in error reduction mask a potential long-term cost to human cognitive and creative faculties.

Similarly, AI systems integrated into decision-making processes, whether in professional settings or personal life, warrant long-term scrutiny. Imagine an AI recommendation engine for financial investments. Initially, it might provide highly accurate and profitable suggestions, leading to user confidence and increased engagement. Short-term metrics would showcase portfolio growth. However, a longitudinal perspective would be essential to observe how users' own financial acumen evolves (or devolves) under the AI's guidance. Do they cease to perform their own due diligence, trusting the AI implicitly? If the AI's underlying algorithms change, or if unforeseen market conditions arise that the AI is not equipped to handle, users who have lost their independent analytical skills could be exposed to significant risk. The long-term impact isn't just about the AI's performance, but about how that performance shapes the user's ongoing capacity for critical assessment and independent judgment. This shift from active decision-maker to passive recipient of AI-generated choices can have profound implications for personal autonomy and resilience.

The societal implications of widely adopted AI also

demand long-term observation. AI systems that influence social interactions, information consumption, or even the formation of opinions can have subtle yet pervasive effects that are not immediately obvious. Consider the impact of AI-driven content personalization on social media platforms. While designed to increase user engagement by showing them content they are likely to enjoy, a longitudinal view might reveal the exacerbation of filter bubbles and echo chambers. Over time, users might be exposed to increasingly narrow perspectives, reinforcing existing beliefs and diminishing their exposure to diverse viewpoints. This can lead to increased polarization, reduced empathy, and a fragmented understanding of societal issues. Evaluating this solely on short-term engagement metrics – likes, shares, time spent on platform – would miss the critical long-term societal cost of diminished civic discourse and understanding. Longitudinal studies are needed to track how these systems shape collective understanding, political discourse, and social cohesion over years.

The development of dependency is another critical area where longitudinal studies are indispensable. Many AI systems are designed for convenience, aiming to simplify tasks and reduce cognitive load. While this can be beneficial, it also carries the risk of fostering over-dependence. For example, AI navigation systems, while immensely helpful, can lead to a decline in users' innate spatial reasoning and map-reading skills. Over time, a person who has exclusively relied on GPS might struggle to navigate unfamiliar environments without it, or even to mentally orient themselves when their device is unavailable. A longitudinal study could track this gradual erosion of a fundamental human skill, highlighting the trade-off between immediate convenience and long-term cognitive capability. This is not to argue against the utility of such systems, but to emphasize the need for awareness and proactive strategies to mitigate potential downsides, such as periodic "unplugged" challenges or educational components that reinforce underlying skills.

Furthermore, the evolution of AI systems themselves, and their interaction with an evolving user base, presents a complex dynamic that necessitates long-term tracking. AI models are often updated, retrained, or have their algorithms refined. User populations also change, with new generations of users entering the digital landscape with different expectations and prior experiences. A longitudinal study can capture how a system's initial impact changes as it matures and as its user base diversifies and adapts. It can reveal whether the AI maintains its intended positive effects, whether unintended consequences emerge over time, or whether the AI's effectiveness diminishes as users become more sophisticated or as the environment in which it operates shifts. This continuous observation allows for iterative refinement and a more adaptive approach to AI governance.

The challenges associated with conducting longitudinal studies are significant, not least of which is the logistical and financial investment required. Maintaining contact with participants over extended periods, ensuring data integrity across different time points, and analyzing complex, evolving datasets demand robust research infrastructure and sustained commitment. Ethical considerations also become more intricate. As participants become more accustomed to an AI system, their consent to ongoing data collection needs to be regularly re-evaluated, ensuring they remain fully informed and comfortable with the process. The potential for revealing long-term negative impacts on individuals or society also raises ethical questions about intervention and responsibility.

However, the importance of these studies cannot be overstated. They provide a vital counterbalance to the often-optimistic narratives surrounding AI deployment, which are frequently based on short-term performance gains. By shifting the focus from immediate utility to sustained impact, we can foster a more responsible and human-centered approach to AI development and integration. This involves asking not just "Does this AI work well now?" but "How will this AI shape human capabilities, behaviors, and societal dynamics in the months and years to come?" This deeper inquiry is essential for building AI systems that truly augment human potential and contribute positively to our future, rather than inadvertently undermining it. The insights gleaned from longitudinal studies are not merely academic curiosities; they are foundational to developing trust, ensuring safety, and realizing the ultimate promise of artificial intelligence in a way that respects and preserves human well-being and societal flourishing.

The very definition of "success" for an AI system must evolve when viewed through a longitudinal lens. Short-term success might be measured by task completion speed, accuracy, or user satisfaction scores gathered in initial weeks. However, long-term success should encompass the sustained development of human skills, the avoidance of detrimental dependencies, the preservation of critical thinking, and the contribution to a healthy and informed society. An AI that achieves high short-term engagement but leads to significant skill degradation or societal fragmentation cannot, by any meaningful measure, be considered truly successful in the long run. Longitudinal studies provide the framework for this more comprehensive assessment, enabling us to differentiate between fleeting gains and enduring benefits.

Moreover, understanding the evolving relationship between users and AI over time is crucial for building adaptive and resilient AI systems. For instance, imagine an AI tutor. In the initial phase, it might excel at introducing new concepts and providing immediate feedback. However, as a student progresses, the AI's role might need to shift from direct instruction to facilitating independent learning, critical inquiry, and problem-solving. A longitudinal study could reveal when and how this transition should occur, identifying points at which the AI's current approach becomes counterproductive. It might show that excessive hand-holding, while beneficial early on, hinders the development of self-directed learning later in a student's educational journey. This suggests that AI systems themselves may need to adapt their strategies and interfaces over time, guided by an understanding of their users' developmental trajectories.

The concept of "AI fluency" is another domain where longitudinal studies are paramount. Just as individuals develop literacy in reading and writing over years, users develop fluency in interacting with and leveraging AI systems. This fluency involves not just knowing how to operate a system, but understanding its limitations, knowing when to trust its outputs, and recognizing when to seek human judgment or alternative approaches. Short-term studies might capture basic operational proficiency, but they miss the deeper understanding that comes with prolonged engagement. A longitudinal perspective can track the development of AI fluency, identifying the factors that foster it and the potential pitfalls that hinder it. This is crucial for ensuring that users can effectively and safely integrate AI into their lives and work, rather than being overwhelmed or misled by it.

In essence, the argument for longitudinal studies is an argument for a more mature and responsible approach to AI evaluation. It acknowledges that the impact of AI is not a static phenomenon, but a dynamic process that unfolds over time, affecting individuals and societies in ways that are often subtle, cumulative, and profound. By investing in the methodologies that allow us to observe these long-term effects, we equip ourselves with the knowledge necessary to guide AI development towards outcomes that are truly beneficial for humanity, avoiding the unintended consequences that might otherwise arise from a myopic focus on short-term gains. This long-term vision is not merely a desirable addition to AI assessment; it is becoming an essential requirement for navigating the complex and rapidly evolving landscape of artificial intelligence.

Tracking user behavior over extended periods is a cornerstone of understanding the long-term impact of AI systems, moving beyond ephemeral impressions to grasp the evolving relationship between humans and artificial intelligence. This involves employing a suite of methodologies designed to capture the nuances of interaction as AI becomes integrated into daily routines, professional workflows, and personal decision-making processes. Unlike snapshots of initial engagement or performance under controlled conditions, these longitudinal approaches reveal the subtle shifts in user reliance, adaptation, and the emergence of both beneficial and detrimental patterns.

One of the most ubiquitous and powerful tools for tracking user behavior over time is in-app analytics. These systems, embedded within the AI application itself, continuously collect data on how users interact with the software. This can range from basic metrics like login frequency, session duration, and feature usage to more granular details such as the specific commands issued, the frequency with which AI-generated suggestions are accepted or rejected, and the time taken to complete tasks with and without AI assistance. For instance, a calendar AI designed to optimize scheduling could log how many meetings a user proposes versus how many the AI suggests, the number of conflicts the AI resolves automatically versus those requiring manual intervention, and the extent to which users customize their scheduling preferences over time. By aggregating this data across a large user base, researchers can identify macro-level trends. They might observe a gradual increase in reliance on AI-driven conflict resolution as users gain confidence in its accuracy, or conversely, a decline in usage of a particular AI feature if it proves cumbersome or less effective than anticipated compared to manual methods. The key lies in the temporal dimension: observing these changes not as isolated events, but as part of a continuous flow, allowing for the identification of inflection points where user behavior shifts significantly. This could be triggered by an AI update, a change in the user's personal circumstances, or simply the natural progression of habit formation.

However, in-app analytics alone can sometimes lack the rich contextual understanding of *why* users behave in certain ways. This is where qualitative methods, such as diary studies, become invaluable. In a diary study, participants are asked to regularly record their experiences, thoughts, and feelings related to using an AI system over a defined period, which could be weeks, months, or even a year.

These entries can be made daily, weekly, or at specific junctures. For a language learning AI, a participant might be prompted to log how they felt using the AI's conversational practice feature after a difficult day, noting whether they found it encouraging or frustrating. They might also record instances where the AI's feedback helped them understand a grammatical nuance they had previously struggled with, or conversely, instances where the AI's corrections felt overly pedantic or missed the intended meaning. These personal narratives provide crucial insights into the emotional and cognitive impact of the AI, shedding light on user satisfaction, perceived value, and the development of personal strategies for using the AI effectively. For example, a participant might report initially relying heavily on the AI's translation feature for every sentence but gradually starting to attempt sentences independently, only turning to the AI for confirmation or for more complex phrases. This progression, captured in their own words, paints a vivid picture of their learning journey and the AI's role within it.

Periodic user surveys, administered at regular intervals throughout the study, offer another layer of data collection. These surveys can be designed to capture specific attitudes, perceptions, and self-reported behavioral changes that might not be evident in passive analytics or diary entries. For instance, a survey administered monthly to users of an AI-powered mental wellness app could ask about their perceived level of stress, their confidence in managing their emotions, and their satisfaction with the app's coping mechanisms.

Over time, the researchers could track trends in these self-reported metrics, correlating them with the app's usage patterns. A user who initially reported high stress and frequent app usage might, over several months, report a decrease in stress levels and a more selective, but still effective, use of the app's tools. This suggests that the AI is not just providing temporary relief but is contributing to a sustainable improvement in the user's well-being. Surveys can also probe for the emergence of specific issues, such as the user's perception of over-reliance, concerns about privacy, or the development of new skills. For example, a user might initially rate their AI system's helpfulness highly, but in a later survey, express concern about forgetting how to perform a task manually, indicating a potential downside that needs to be addressed.

The formation of habits is a critical aspect of long-term AI interaction that these longitudinal methods help to illuminate. Humans are creatures of habit, and AI systems, particularly those designed for convenience and efficiency, are potent catalysts for habit formation. Consider an AI that manages a user's email inbox, prioritizing messages, drafting replies, and archiving routine correspondence. Initially, the user might meticulously review every AI-generated draft and every categorization decision. Over weeks and months, however, as the AI consistently demonstrates accuracy and efficiency, the user may develop a habit of approving AI-generated actions with minimal review, or even allowing the AI to act autonomously on certain types of communication. This shift from active oversight to passive trust represents a significant change in user behavior. Longitudinal analytics can track the decreasing time spent reviewing AI outputs, the increasing rate of automated replies, and the declining frequency of manual email sorting. Diary entries might reveal the user's thought process: a growing confidence in the AI's judgment, a desire to save time, or even a subtle feeling of impatience when manual tasks interrupt their workflow. Identifying these habit formation patterns is crucial for understanding user engagement and for designing AI systems that foster beneficial habits without creating undesirable dependencies.

Moreover, longitudinal studies are essential for detecting emergent issues that are simply not visible in short-term evaluations. A usability test might reveal that users can easily initiate a task with an AI assistant. However, a longitudinal study could uncover that after several months, users begin to experience "interface fatigue" with that particular AI, finding its prompts repetitive or its workflow unnecessarily complex for routine actions. This could lead to a gradual decrease in engagement or the development of workarounds that bypass the AI's intended design. Similarly, an AI designed for personalized learning might show excellent initial results in improving test scores. But over a longer period, it might inadvertently create "performance gaps" if it consistently focuses on areas where the student excels, neglecting more challenging subjects that require deeper, more sustained human guidance. This would only become apparent when observing the student's overall academic progress and skill development across a broader curriculum over an academic year.

The concept of "skill degradation" is a prime example of an emergent issue that longitudinal tracking can expose. If an AI consistently performs complex calculations, generates reports, or even diagnoses medical conditions, users may gradually cede these cognitive tasks, leading to a diminishment of their own skills in these areas. For instance, pilots using advanced AI co-pilots might, over years, experience a reduction in their manual flying skills or their ability to intuitively assess atmospheric conditions. While the AI ensures safety in the present, the long-term consequence could be a generation of pilots less equipped to handle situations where the AI is unavailable or malfunctions. Longitudinal studies, through a combination of analytics (tracking the frequency of manual overrides versus AI actions) and periodic skill assessments (e.g., simulated flight scenarios without AI assistance), can quantify this degradation and highlight the need for ongoing training or human oversight to maintain critical human competencies.

Furthermore, longitudinal tracking allows researchers to understand how users adapt their own behaviors and strategies in response to an evolving AI. AI systems are not static; they are frequently updated, retrained, and refined. As an AI system improves or changes, user behavior patterns will likely shift in response. An initial

AI writing assistant might be praised for its ability to correct basic grammatical errors. Later versions might offer stylistic suggestions, sentence rephrasing, or even content generation. Users will need to adapt their interaction strategies accordingly. A longitudinal study could track how users' initial reliance on simple error correction evolves into a more sophisticated engagement with stylistic refinement or content ideation. Diary studies might reveal users experimenting with different prompts to elicit more creative output from the AI, or conversely, learning to ignore certain types of AI suggestions that they find intrusive or unhelpful. This dynamic interplay between AI evolution and user adaptation is a complex phenomenon that can only be understood through sustained observation.

The ethical implications of AI also come into sharper focus with longitudinal studies. Issues of privacy, bias, and autonomy can manifest or intensify over time. For example, an AI that collects extensive personal data for personalized recommendations might raise minor privacy concerns initially. However, as it accrues years of data, the potential for misuse, data breaches, or invasive profiling increases significantly. Longitudinal studies can track user sentiment regarding privacy over time, correlating it with the AI's data collection practices and security measures. Similarly, subtle biases embedded in an AI's algorithms might not be apparent in short-term interactions but could lead to discriminatory outcomes when the AI is used consistently for critical decisions, such as loan applications or job candidate screening. Tracking the outcomes of these decisions over extended periods can reveal patterns of systemic bias that might otherwise go unnoticed. The evolution of user autonomy is another critical aspect. If an AI consistently makes decisions for the user, whether it's what news to read, what route to take, or what product to buy, it can gradually erode the user's capacity for independent decision-making. Longitudinal studies can observe this gradual shift, perhaps through surveys that measure perceived control and self-efficacy, or through qualitative interviews that explore how users experience their decision-making process when influenced by AI.

In practice, combining these methodologies is often the most effective approach. In-app analytics provide the broad strokes of user behavior, identifying trends and anomalies across a large population. Diary studies offer deep, qualitative insights into the user experience, explaining the "why" behind the data. Periodic surveys act as a bridge, allowing for the collection of quantitative attitude and perception data at regular intervals, helping to triangulate findings and assess progress towards long-term goals or the emergence of potential problems. For instance, researchers studying an AI that assists with financial planning might use analytics to see how often users accept AI-generated investment advice. They might then use diary entries to understand if users feel more or less financially literate, and employ surveys to gauge their confidence in managing their investments independently. Over a year, they might observe that while AI-driven portfolio growth is consistent, users report a decreasing sense of financial agency, prompting a redesign that empowers users to understand the AI's reasoning and participate more actively in decision-making.

The implementation of such longitudinal tracking requires careful consideration of ethical guidelines and participant consent. As users spend more time with an AI system, their understanding of its capabilities and limitations deepens, and their reliance may increase. This necessitates a continuous process of informed consent, where participants are regularly reminded of the data being collected, its purpose, and their right to withdraw. Transparency about how data is used, stored, and protected is paramount, especially as the volume of accumulated data grows. Furthermore, researchers must be prepared for the possibility that longitudinal studies may reveal significant long-term negative impacts, raising questions about their responsibility to intervene, report findings, or advocate for system redesign. The commitment to longitudinal study is therefore not just a methodological choice, but an ethical imperative to fully understand and responsibly manage the profound and evolving influence of AI on human lives. By diligently tracking user behavior over time, we move from a superficial understanding of AI's immediate utility to a profound grasp of its enduring influence, guiding us towards a future where AI genuinely augments human potential without undermining essential human capacities or societal well-being.

The persistent engagement with artificial intelligence systems over extended durations can precipitate subtle yet significant alterations in human cognitive functions and behavioral patterns. These shifts, often imperceptible in short-term interactions, become salient when observed through the lens of longitudinal studies, offering critical insights into the developing symbiosis between humans and intelligent machines. One primary area of concern is the potential impact on **critical thinking skills**. As AI systems become adept at processing information, synthesizing data, and even generating solutions, individuals may develop a reliance that bypasses the deep cognitive effort typically associated with critical analysis. For instance, a student consistently using an AI to summarize complex academic texts might, over time, experience a diminished capacity to engage in independent critical appraisal, to identify underlying assumptions, or to formulate nuanced arguments based on their own interpretation. The AI's distilled summaries, while efficient, can inadvertently short-circuit the learner's development of these crucial intellectual faculties. Longitudinal tracking, through periodic assessments of analytical essay writing, debate participation, or even problem-solving tasks presented without AI assistance, can help quantify this potential decline. Researchers might observe a decrease in the complexity of student-generated arguments or an increased tendency to accept AI-generated conclusions without interrogation.

Closely related is the potential for **memory recall and information retrieval** to be affected. When AI tools readily provide answers, recall specific facts, or organize vast amounts of personal data, the need for internal memory consolidation and retrieval mechanisms may diminish. Consider an individual using an AI-powered personal assistant that meticulously logs all past conversations, appointments, and project details. While immensely convenient, this persistent externalization of memory could lead to a gradual weakening of the brain's own ability to recall information spontaneously. Over months or years, this might manifest as a decreased ability to access stored knowledge without prompting, or a reliance on the AI to trigger memories that were once readily accessible. Diary studies could capture qualitative data where participants express frustration when unable to recall something they "should know" without consulting their AI. Furthermore, surveys measuring self-reported memory performance and objective cognitive tests administered at regular intervals can provide quantitative evidence of any such degradation. The challenge lies in distinguishing genuine cognitive decline from a simply more efficient allocation of cognitive resources, where the brain opts for external aids when available.

The
autonomy of decision-making is another profound aspect that longitudinal studies can illuminate. AI systems, especially those designed for personalization and optimization, often nudge users towards specific choices. Think of AI recommendation engines for news, entertainment, or even professional development. While these recommendations can be beneficial, a consistent pattern of accepting AI-driven suggestions

without independent deliberation can gradually erode a user's sense of agency. Over an extended period, individuals might find themselves less inclined to explore alternative options or to question the AI's rationale, passively following a pre-determined path. This can be particularly concerning in domains requiring complex judgment, such as financial planning or medical treatment choices. Longitudinal research might involve observing decision-making processes in simulated environments or through in-depth interviews that explore users' perceived control and confidence in their choices. A notable example could be observing how users' engagement with personalized learning platforms evolves. Initially, users might actively choose courses based on their interests. Over time, if the AI consistently curates their learning path based on perceived future needs or performance metrics, users might simply follow the AI's directive, potentially missing out on avenues for serendipitous discovery or personal exploration. This subtle shift from active selection to passive acceptance can have long-term consequences for intellectual curiosity and self-directed growth.

Beyond individual cognitive processes, AI's long-term impact on
social interactions also warrants careful scrutiny. AI-driven communication tools, social media algorithms, and even virtual companions can reshape how humans connect with each other. For instance, the increasing use of AI-generated text for initial drafts of emails or social media posts could lead to a homogenization of communication styles, reducing the expression of

individual personality and nuance. If AI filters conversations or suggests responses, the spontaneous, often messy, but authentic exchange of ideas might be supplanted by a more curated and predictable form of interaction. Longitudinal studies observing communication patterns, sentiment analysis of user-generated content over time, and surveys on perceived social connection quality can reveal these trends. One might observe a decline in the use of idiomatic expressions or unique personal anecdotes in favor of more generic, AI-vetted phrases. This could, in turn, impact the depth of interpersonal relationships, as genuine connection often thrives on authentic self-expression.

The concept of **competency and skill maintenance** is another critical area affected by long-term AI interaction. When AI automates complex tasks, human proficiency in those tasks may atrophy. A classic example is the reliance on GPS navigation systems. While incredibly useful, consistent use of AI navigation can potentially weaken an individual's innate spatial reasoning and sense of direction. Over years of relying on turn-by-turn directions, a person might find it increasingly difficult to navigate unfamiliar areas without technological assistance, or even to estimate distances and travel times accurately. Longitudinal studies involving periodic map-reading tests, navigation challenges in unfamiliar environments without AI assistance, and self-reported confidence in independent navigation can help quantify this effect. Similarly, professionals using AI for complex data analysis or report generation might, over time, experience a decline in their ability to perform these tasks manually, leading to a vulnerability if AI systems are unavailable or their outputs are suspect. The challenge for researchers is to distinguish between the natural evolution of skill acquisition, where humans learn to leverage tools effectively, and the detrimental loss of fundamental competencies due to over-reliance.

Furthermore, the **formation of new cognitive habits and mental models** is an inevitable consequence of prolonged AI engagement. AI systems, by their very nature, encourage users to adopt specific ways of thinking and interacting with information. For instance, an AI that categorizes emails automatically trains users to accept its classifications, rather than developing their own system for prioritizing or filing. Over time, this can lead to a mental model where information organization is seen as an external, automated process rather than an internal cognitive skill. Similarly, AI tools that offer predictive text or sentence completion can shape how individuals construct their thoughts, potentially leading to more formulaic sentence structures and a reduced capacity for spontaneous creative writing. Longitudinal studies can track the evolution of these mental models through think-aloud protocols, user interviews exploring their problem-solving strategies, and analysis of the structure and creativity in user-generated content over time. For example, a writer might initially be surprised by AI-generated sentence suggestions but, over months, start unconsciously adopting similar phrasing in their own original writing, reflecting a shift in their internal writing process.

The

development of trust and skepticism in AI systems is another dynamic process best understood through longitudinal observation. Initial interactions might be marked by a mixture of curiosity, excitement, and caution. As users gain experience, their trust can either solidify or erode based on the AI's performance, transparency, and perceived fairness. This evolving trust landscape is critical. An AI that consistently delivers accurate and helpful results might foster deep reliance, while an AI that exhibits biases, errors, or a lack of transparency could lead to escalating skepticism and eventual rejection. Longitudinal studies can gauge these attitudes through regular surveys measuring user trust, perceived reliability, and satisfaction. Qualitative data from diary studies can provide rich narratives about the moments that shaped this trust, such as a critical error that caused significant inconvenience, or an instance where the AI provided an unexpectedly insightful solution. For example, users of an AI-powered financial advisory service might initially treat every suggestion with suspicion. If the AI consistently helps them achieve their financial goals and explains its reasoning clearly, their trust might grow, leading them to delegate more complex financial decisions. Conversely, a single instance of demonstrably poor advice could shatter that trust, leading them to revert to manual methods or seek human advice.

The impact of AI on

human motivation and engagement is also a key area for longitudinal study. AI systems designed to enhance productivity or provide personalized experiences can, over time, alter how individuals perceive effort, challenge, and reward. For example, an AI that automates many of the tedious aspects of a task might reduce the user's intrinsic motivation if the challenge and effort are removed, leading to boredom or a feeling of disengagement. Conversely, an AI that gamifies tasks or provides constant positive reinforcement could foster sustained engagement. Longitudinal research, employing methods such as tracking task completion rates, measuring subjective motivation levels through questionnaires, and observing the variety of tasks users undertake, can reveal these effects. A study of an AI-powered fitness tracker, for instance, might initially show high engagement as users strive to meet AI-set goals. Over several months, however, some users might become demotivated if the AI's challenges become too easy or too difficult, or if the AI's feedback becomes repetitive, leading to a decline in usage. Understanding these dynamics is crucial for designing AI that sustains, rather than saps, human motivation.

Finally, longitudinal studies are paramount for understanding the
cumulative effects of AI exposure on well-being and mental health. While an AI might offer immediate benefits, its continuous presence and influence could have unforeseen long-term psychological consequences. For instance, an AI designed to manage social media feeds and content exposure might, over time, contribute to a distorted perception of reality or exacerbate

feelings of social comparison and inadequacy if its algorithms are not carefully designed to promote balanced perspectives. Similarly, AI systems that engage users in prolonged, intense interactions without adequate breaks or transitions could contribute to mental fatigue or even addictive behaviors.

Longitudinal research incorporating mental health assessments, surveys on perceived stress and anxiety, and qualitative interviews exploring users' subjective experiences of their mental state in relation to AI use is essential. For example, a user who relies heavily on an AI for companionship might, over years, experience a gradual erosion of their social skills with other humans, leading to increased isolation and dependence on the AI. Such complex, long-term effects can only be responsibly understood and addressed through sustained, rigorous investigation. The ethical imperative here is to ensure that AI development prioritizes human flourishing by actively investigating and mitigating any potential long-term detriments to cognitive function, autonomy, social connection, and overall well-being.

The pervasive integration of artificial intelligence into the fabric of daily life and professional endeavors transcends individual cognitive and behavioral shifts, ushering in profound societal and ethical transformations. Longitudinal studies, therefore, must extend their purview beyond the individual user to encompass the macro-level impacts that AI deployment has on industries, social structures, and the very ethical landscapes within which we operate. This extended examination is not merely an academic exercise; it is a critical imperative for navigating the complex future shaped by increasingly sophisticated AI.

One significant area of concern lies in the **transformative effects on industries and labor markets**. As AI systems become more capable of performing tasks previously exclusive to human expertise, entire sectors may undergo radical restructuring. Consider the impact on healthcare. AI-powered diagnostic tools, robotic surgery assistants, and personalized treatment planning systems, while promising unprecedented efficiency and accuracy, also raise questions about the future roles of physicians, nurses, and technicians. Longitudinal tracking of healthcare systems that adopt these technologies can reveal not just improvements in patient outcomes, but also shifts in workforce composition, the emergence of new specialized roles (e.g., AI-assisted diagnostics interpreters), and the potential displacement of workers whose skills are rendered obsolete. Studies might track the demand for certain medical specializations over a decade, correlating it with the adoption rates of AI diagnostic tools. Furthermore, the economic implications are vast. Will AI-driven productivity gains lead to widespread prosperity, or will they exacerbate existing inequalities, concentrating wealth and power in the hands of those who control and deploy these advanced technologies? Monitoring the evolution of income distribution, employment rates in AI-augmented versus traditional sectors, and the accessibility of AI-driven services over extended periods is crucial for understanding these complex socio-economic dynamics. The rise of AI-powered customer service, for instance, might initially lead to cost savings for businesses, but over years, it could significantly alter consumer expectations for responsiveness and personalization, while simultaneously reducing the need for human customer service representatives, potentially leading to large-scale job displacement. Longitudinal surveys of displaced workers, coupled with economic modeling of AI adoption rates across various industries, can provide a stark picture of these shifts.

Beyond the economic sphere, AI's long-term influence on **social structures and community dynamics** warrants deep investigation. AI algorithms already play a significant role in shaping what information we consume, who we connect with online, and even how we perceive social norms. Consider the persistent exposure to algorithmically curated news feeds. Over years, this can lead to the formation of distinct echo chambers, reinforcing existing beliefs and potentially contributing to increased societal polarization. Longitudinal content analysis of social media platforms, combined with demographic surveys measuring political attitudes and social cohesion, could reveal how prolonged AI-driven content delivery influences public discourse and interpersonal trust. Furthermore, the increasing sophistication of AI companions and virtual assistants could alter the nature of human relationships. If individuals increasingly turn to AI for emotional support, companionship, or even intellectual stimulation, what are the long-term implications for human-to-human interaction, empathy, and the formation of deep social bonds? Studies might involve in-depth qualitative research with individuals who have maintained long-term relationships with AI companions, exploring their subjective experiences of connection, loneliness, and social development over time. Observing the patterns of social interaction within communities that have high penetration of AI-driven social platforms or pervasive AI assistants could also reveal subtle shifts in community engagement, civic participation, and the nature of shared experiences. For example, a longitudinal study of a town where smart home AI is universally adopted might reveal changes in inter-neighborly communication, reliance on local community groups versus AI-mediated services for assistance, and the overall sense of collective identity.

The development and deployment of AI also give rise to novel and complex **ethical dilemmas that mature over time**. As AI systems become more autonomous and their decision-making processes more opaque, questions of accountability, fairness, and bias become increasingly pressing. For instance, in the realm of criminal justice, AI systems used for risk assessment in sentencing or parole decisions, if biased, can perpetuate and amplify existing societal inequities. Longitudinal studies tracking the outcomes of individuals processed through such systems, alongside rigorous auditing of the AI's decision-making patterns over years, are essential for identifying and rectifying systemic biases. This involves not just looking at individual erroneous decisions, but at the aggregate impact on specific demographic groups over extended periods. The challenge lies in the dynamic nature of AI; as models are retrained and updated, their underlying biases can shift, making continuous monitoring a necessity.

Furthermore, the increasing integration of AI into sensitive domains like healthcare, finance, and defense raises profound questions about **data privacy and security in the long run**. As AI systems ingest and process vast amounts of personal data to provide their services, the potential for breaches, misuse, or unintended surveillance grows. Longitudinal studies examining the evolution of data protection regulations, user awareness of data usage, and the frequency and impact of AI-related data security incidents are critical. How do individuals' perceptions of privacy change as they become more accustomed to sharing data with AI systems over a decade? Do they become more complacent, or do high-profile breaches lead to increased vigilance and demand for stronger protections? Analyzing trends in cybersecurity threats targeting AI systems and the effectiveness of mitigation strategies over extended periods will be crucial. The implications extend to national security, where AI-powered autonomous weapons systems present unprecedented ethical quandaries regarding human control, accountability in warfare, and the potential for unintended escalation. Long-term international dialogues and the establishment of robust ethical frameworks for the development and deployment of such technologies are paramount.

To effectively monitor these broad societal and ethical

ramifications, a suite of methodological approaches is required, extending beyond individual-level data collection.

Trend analysis of aggregate data is fundamental. This involves tracking key societal indicators—such as employment figures by sector, income inequality metrics, measures of social trust, crime rates, and public health statistics—and analyzing their correlation with the widespread adoption of AI technologies over time. For example, by mapping the increase in AI adoption in logistics and transportation against trends in truck driver employment, a clearer picture of AI's impact on that workforce can emerge. This requires access to robust, long-term national and international datasets.

Crucially,

ethical impact assessments (EIAs), when conducted longitudinally, offer a proactive framework for anticipating and mitigating potential harms. Unlike one-off assessments, longitudinal EIAs would involve repeated evaluations of AI systems and their deployment contexts at defined intervals—say, every two to five years. These assessments would revisit initial predictions, incorporate actual observed impacts, and revise strategies for responsible development and governance. For instance, an EIA for an AI-powered hiring tool initially might focus on algorithmic bias. A longitudinal EIA, conducted after several years of the tool's use, would examine not only persistent algorithmic bias but also its impact on workforce diversity trajectories, employee retention rates, and the evolution of hiring practices and organizational culture.

Such assessments should be iterative, adapting to new insights and unforeseen consequences.

Furthermore,
public discourse analysis provides an invaluable lens through which to understand evolving societal attitudes, concerns, and ethical norms surrounding AI. By systematically analyzing news media, academic publications, policy debates, and social media conversations over extended periods, researchers can identify emerging ethical challenges, public perception shifts, and areas of consensus or contention. For example, tracking the language and framing used in public discussions about AI in healthcare over a decade could reveal a gradual shift from initial excitement about technological advancement to a more nuanced debate about patient autonomy, data rights, and the equitable distribution of AI-driven medical benefits. Sophisticated natural language processing techniques can automate the analysis of vast quantities of textual data, identifying recurring themes, sentiment shifts, and the emergence of new ethical terminology related to AI.

Finally, fostering
interdisciplinary collaboration and ongoing dialogue among technologists, ethicists, social scientists, policymakers, and the public is not a method in itself, but a foundational requirement for effective long-term monitoring. No single discipline possesses all the necessary insights or tools to grasp the full spectrum of AI's societal and ethical implications. By bringing diverse perspectives together, longitudinal studies can be designed more comprehensively, their

findings interpreted more robustly, and actionable recommendations developed that address the multifaceted challenges of AI's enduring presence. This collaborative approach ensures that the focus remains on developing and deploying AI in ways that genuinely benefit humanity, promoting flourishing rather than inadvertently undermining it. The commitment to these ongoing, holistic evaluations is what will truly allow us to harness the transformative potential of AI while remaining vigilant against its potential pitfalls, ensuring a future where technological advancement aligns with our deepest human values.

The integration of artificial intelligence into educational environments presents a fertile ground for longitudinal study, offering a unique window into how advanced technologies can reshape learning, teaching, and the very development of intellectual capabilities over extended periods. Unlike studies that capture snapshots of AI's impact, a longitudinal approach allows us to witness the gradual evolution of its influence, observing how students and educators adapt, how pedagogical strategies shift, and what enduring consequences emerge. This necessitates tracking individuals and institutions over several years, observing the subtle, cumulative effects of AI-powered tools on learning outcomes, the acquisition of critical skills, and the broader educational ecosystem. Such an endeavor moves beyond simply evaluating the efficacy of a new software or hardware; it delves into the fundamental transformation of the learning experience itself.

Consider a cohort of students entering primary education at a time when AI-driven adaptive learning platforms are becoming increasingly prevalent. These platforms, designed to personalize the pace and content of instruction based on individual student performance, promise to address the diverse needs within a classroom more effectively than traditional one-size-fits-all methods. In the initial years of their schooling, these students might exhibit accelerated progress in foundational subjects like mathematics and reading, as the AI tutors identify and address knowledge gaps with tailored exercises and immediate feedback. Teachers, in turn, might find their roles evolving from direct instructors to facilitators, spending more time on higher-order thinking skills, socio-emotional development, and addressing the unique challenges that AI cannot fully replicate. The AI handles the drudgery of repetitive practice, freeing up human educators for more impactful interactions. This initial phase could be characterized by a palpable enthusiasm for the efficiency and individual attention offered by these tools, with preliminary assessments showing promising gains in standardized test scores.

However, as these students progress through their educational journey, say into secondary school, the long-term implications of this AI-centric learning begin to surface. A crucial area of observation would be the development of problem-solving skills and critical thinking. If students have consistently relied on AI to provide step-by-step solutions or to flag potential errors in their reasoning, they may develop a dependency that hinders their ability to tackle novel, complex problems independently. When faced with a challenge that deviates from the patterns the AI has been trained on, or one that requires abstract reasoning or creative synthesis, these students might struggle. A longitudinal study would track not just their performance on standard assessments, but also their approaches to open-ended projects, their willingness to engage in trial-and-error, and their capacity for independent exploration. Researchers might employ qualitative methods, such as in-depth interviews and observational studies of group work, to gauge the depth of their understanding and their metacognitive skills—their awareness of their own learning processes.

Furthermore, the educator's role continues to transform. While AI might initially be seen as a powerful assistant, its pervasive integration could lead to a de-skilling of certain pedagogical aspects if not managed carefully. Teachers who rely heavily on AI for lesson planning, assessment generation, and even delivering direct instruction might see their own pedagogical expertise atrophy over time. A longitudinal study would examine how professional development needs evolve. Instead of focusing on subject matter delivery, educators might require training in AI literacy, data interpretation for pedagogical insights, and the ethical considerations of using AI with students. The challenge lies in ensuring that AI enhances, rather than replaces, the nuanced judgment and empathic connection that are hallmarks of effective teaching. Over a decade, we might observe two divergent paths: one where educators expertly leverage AI to amplify their teaching capabilities and foster deeper learning, and another where AI becomes a crutch, leading to a more passive and less engaging educational experience for both students and teachers.

The impact on the curriculum itself is another critical dimension for long-term observation. As AI becomes more adept at handling routine tasks and providing factual information, the emphasis in education may naturally shift towards skills that are uniquely human. Creativity, collaboration, communication, emotional intelligence, and ethical reasoning—these are areas where AI currently lags significantly. A longitudinal study would monitor how curricula adapt to prioritize these "21st-century skills." Are schools actively redesigning their programs to cultivate these abilities, or are they merely layering AI tools onto existing structures? The long-term success of AI integration in education will depend on whether it frees up time and resources to focus on these higher-level human competencies, rather than simply automating existing educational processes. Observing student engagement in group projects that require negotiation and consensus-building, or their participation in debates that demand critical argumentation and persuasive communication, can provide valuable insights into this shift.

Beyond cognitive development, the socio-emotional impact of sustained AI interaction in schools is a vital area for longitudinal scrutiny. Students who spend a significant portion of their formative years interacting with AI tutors and personalized learning systems might develop different patterns of social interaction and emotional regulation compared to their predecessors. While AI can offer consistent, unbiased feedback in academic contexts, it cannot fully replicate the complex social and emotional learning that occurs through peer interaction and human mentorship. A long-term study could track the development of empathy, the ability to navigate social conflict, and the formation of meaningful relationships. Are students more adept at understanding and responding to the emotional cues of their peers and teachers if their primary academic interactions are with non-sentient systems? Conversely, could over-reliance on AI for instant gratification and immediate feedback lead to impatience with the slower, more nuanced pace of human relationships? This requires careful, long-range observation of social dynamics within classrooms and schools that have deeply integrated AI.

The equity implications of AI in education also warrant extended examination. While AI-powered tools promise to democratize access to high-quality education by offering personalized support regardless of geographical location or socioeconomic background, there is a significant risk of exacerbating existing disparities. The digital divide, for instance, could widen if access to the most sophisticated and beneficial AI learning tools is limited to well-resourced schools and families. Over a decade, a longitudinal study would need to track the performance and skill development of students across a spectrum of socioeconomic and geographical contexts. Are students in under-resourced schools, who may have access only to basic AI tools or even suffer from unreliable internet connectivity, falling further behind their peers who benefit from cutting-edge AI integration? Furthermore, the development and training data for AI algorithms can carry inherent biases. A longitudinal study could reveal how these biases, if unaddressed, perpetuate and amplify existing inequities in educational outcomes for students from marginalized groups, affecting everything from college admissions to career path recommendations. Continuous auditing of AI algorithms and their impact on diverse student populations over time is essential to ensure fairness.

Moreover, the long-term impact on educators' professional identity and job satisfaction needs careful consideration. As AI takes over more instructional and administrative tasks, teachers may face anxieties about job security or a diminished sense of professional autonomy. Conversely, they might experience increased job satisfaction if AI liberates them from repetitive tasks, allowing them to focus on more engaging and meaningful aspects of teaching, such as mentoring, curriculum development, and fostering a positive classroom culture. A longitudinal study involving surveys, interviews, and performance reviews of educators over several years would provide crucial data on these evolving professional experiences. It would help identify best practices for AI integration that support, rather than undermine, the teaching profession, ensuring that educators remain empowered and motivated in an AI-augmented educational landscape.

The ethical considerations surrounding AI in education are not static; they evolve with the technology and its adoption. Issues such as data privacy, algorithmic transparency, and student consent become increasingly complex as AI systems gather more granular data on student learning behaviors, cognitive patterns, and even emotional responses. A longitudinal study would track how these ethical challenges manifest and are addressed over time. For example, initial concerns might focus on the security of student data. After several years of widespread AI use, the focus might shift to the implications of predictive analytics generated by AI – how do these predictions influence student pathways, and what are the ethical boundaries of such foresight? Do students and parents fully understand how their data is being used, and do they have meaningful control over it? Long-term studies would require ongoing engagement with ethical frameworks, policy development, and public discourse to ensure that AI in education aligns with societal values and protects the rights of learners.

A hypothetical longitudinal study could begin by tracking a cohort of 1,000 students entering middle school, with diverse backgrounds and attending schools with varying levels of AI integration. For the first five years, researchers would collect data on academic performance (grades, standardized test scores), engagement levels (participation in class, completion rates), and basic skill development (reading comprehension, mathematical problem-solving). During this period, schools with higher AI integration might show initial gains in engagement and targeted skill improvement. Educators in these schools would be interviewed periodically to understand their experiences with AI tools, their evolving pedagogical approaches, and their perceptions of student learning.

In the subsequent five years, as these students move into high school and then potentially into early college or vocational training, the focus of the study would broaden. Researchers would look for indicators of higher-order thinking, creativity, and collaborative skills. They might administer tasks that require critical analysis of complex information, problem-solving in novel contexts, and teamwork. Qualitative data would become more prominent, with researchers observing group projects, conducting interviews about students' approaches to challenging tasks, and assessing their capacity for self-directed learning. Simultaneously, the study would examine the long-term impacts on educators, looking at career progression, job satisfaction, and the development of new competencies related to AI. Issues of equity would be a central theme, with researchers analyzing how AI integration has affected students from different socioeconomic and demographic groups, paying close attention to any emerging disparities in outcomes. Data privacy concerns would be revisited, assessing how student and parent awareness and attitudes have evolved, and whether regulatory frameworks have kept pace with technological advancements.

The findings from such a longitudinal case study would be invaluable. They could reveal that while AI offers significant benefits in personalizing instruction and improving efficiency in foundational learning, its long-term success hinges on a deliberate and thoughtful integration that prioritizes the development of uniquely human skills. It might highlight the critical need for ongoing professional development for educators, equipping them not just to use AI tools, but to critically evaluate their impact and adapt their teaching strategies accordingly. The study could also underscore the importance of proactive ethical oversight and robust data governance to ensure that AI in education fosters equity and empowers all learners, rather than inadvertently entrenching existing disadvantages. Ultimately, a sustained, multi-year examination provides the necessary depth to understand how AI truly reshapes the landscape of learning, moving beyond immediate impacts to reveal the enduring legacies it leaves on students, educators, and the educational system as a whole. This perspective is essential for making informed decisions about the future of AI in education, ensuring that it serves as a tool for genuine pedagogical advancement and human flourishing.

Chapter 11: Practical Application: Healthcare Devices

The deployment of artificial intelligence within the domain of medical diagnostics represents a profound shift, promising to augment the capabilities of healthcare professionals and enhance the precision and efficiency of patient care. At the heart of this transformation lies the principle of human-centered AI, a philosophy that prioritizes the needs, capabilities, and well-being of the end-users – in this context, clinicians and, by extension, patients. This approach moves beyond simply developing algorithms that can identify patterns or make predictions; it focuses on creating AI systems that act as trusted collaborators, seamlessly integrated into the complex fabric of clinical practice. The objective is not to replace human expertise, but to amplify it, enabling physicians, radiologists, pathologists, and other specialists to perform their duties with greater accuracy, speed, and confidence, ultimately leading to improved diagnostic outcomes and enhanced patient safety.

One of the most compelling applications of human-centered AI in medical diagnostics lies in the interpretation of medical imaging. The sheer volume and complexity of radiological scans, such as X-rays, CT scans, and MRIs, can present a significant challenge for even the most experienced radiologists. AI algorithms, trained on vast datasets of annotated images, can now identify subtle anomalies that might elude the human eye, flag suspicious regions for closer inspection, and even provide preliminary classifications of potential pathologies. However, the effectiveness of these tools hinges on their human-centered design. This means that the AI should not operate as a black box, presenting a diagnosis without explanation. Instead, it must provide clear, interpretable insights. For instance, an AI system designed to detect lung nodules on a CT scan should not merely highlight a suspicious area; it should indicate *why* it flagged that area, perhaps by pointing to specific textural characteristics, shape deviations, or density variations that are indicative of abnormality. This transparency allows the radiologist to critically evaluate the AI's suggestion, compare it with their own interpretation, and make a more informed final decision. The AI becomes a sophisticated co-pilot, offering data-driven perspectives that complement the clinician's experience and judgment.

Consider a scenario where an AI system analyzes a mammogram for signs of breast cancer. A human-centered design would ensure that the AI not only identifies potential malignant masses but also quantifies their likelihood of being cancerous, provides measurements of their size and location, and perhaps even highlights surrounding tissue changes that might be relevant. Crucially, the interface would present this information in a way that is easily digestible and actionable within the existing mammography workflow. Instead of overwhelming the radiologist with raw data, the AI would present a summarized report, visually highlighting areas of concern on the image itself, and providing a confidence score for its findings. This allows the radiologist to quickly focus their attention on the most critical areas, potentially reducing the time spent on each scan and increasing the overall throughput of the radiology department. Furthermore, the AI could be designed to learn from the radiologist's feedback. If a radiologist consistently disagrees with a particular type of AI recommendation, the system should be able to adapt and refine its future analyses, thereby fostering a collaborative and iterative learning process between human and machine.

Beyond image analysis, human-centered AI is revolutionizing the interpretation of other forms of patient data, such as electronic health records (EHRs), laboratory results, and even genomic information. AI can sift through enormous volumes of data to identify subtle patterns that might indicate an increased risk of developing certain diseases, predict a patient's response to specific treatments, or alert clinicians to potential adverse drug interactions. The human-centered aspect here is critical for ensuring that these insights are presented in a clinically relevant and actionable manner. An AI predicting a patient's risk of sepsis, for example, should not just output a percentage. It should identify the specific clinical indicators that contributed to that prediction – such as rising white blood cell count, increased heart rate, and specific vital sign trends – and suggest potential next steps, such as ordering further tests or initiating early interventions. This empowers the clinician with context, enabling them to understand the AI's reasoning and integrate its recommendations into their clinical decision-making process.

The integration of AI into clinical workflows is another

paramount consideration for a human-centered approach. A powerful diagnostic AI is of little use if it disrupts the established routines and processes of a busy hospital or clinic. Therefore, these systems must be designed to integrate seamlessly into existing IT infrastructure and user interfaces. This could involve embedding AI-powered diagnostic assistance directly within the EHR system, so that clinicians can access these tools without having to switch between multiple applications. For example, when a physician is reviewing a patient's chart, the AI could automatically analyze newly entered data or review past diagnostic reports, flagging any potential concerns or suggesting relevant diagnostic pathways. The user interface should be intuitive and familiar, minimizing the learning curve for healthcare professionals and reducing the cognitive load associated with adopting new technologies. The goal is to make the AI an invisible, yet invaluable, assistant that enhances, rather than complicates, the daily work of clinicians.

Maintaining clinician oversight and control is a cornerstone of human-centered AI in medical diagnostics. The AI should always be viewed as a tool to support, not supplant, the judgment of the healthcare professional. This means that the final diagnostic decision must always rest with the clinician. The AI's role is to provide information, highlight potential issues, and offer probabilities, but it should not be empowered to make independent diagnoses. This principle is essential for several reasons. Firstly, it ensures accountability; the clinician remains responsible for the patient's care. Secondly, it leverages the invaluable experience, intuition, and holistic understanding that human clinicians possess, which AI, at present, cannot fully replicate. For instance, an AI might identify a statistically significant anomaly in an X-ray, but a seasoned radiologist can often discern from subtle contextual clues – such as the patient's age, medical history, and the specific angle of the image – whether that anomaly is clinically significant or benign.

To facilitate this oversight, AI systems should be designed with clear mechanisms for clinicians to override or modify AI suggestions. If an AI flags a particular lesion as potentially malignant, the radiologist must have the ability to confidently label it as benign based on their expertise, and the AI system should record and learn from this interaction. This feedback loop is crucial for refining the AI's performance over time and ensuring that it remains aligned with clinical realities. Furthermore, the AI should be transparent about its limitations. It should clearly indicate when its confidence in a particular finding is low or when the data it has analyzed is incomplete or of poor quality. This honesty about its capabilities builds trust and prevents over-reliance on potentially flawed AI outputs.

The ethical implications of using AI in medical diagnostics are also deeply intertwined with the human-centered approach. Ensuring fairness and equity is paramount. AI algorithms are trained on data, and if that data reflects existing biases in healthcare – such as disparities in diagnosis or treatment based on race, gender, or socioeconomic status – the AI can inadvertently perpetuate or even amplify these biases. A human-centered design process therefore necessitates rigorous attention to data diversity and bias detection. Developers must actively seek out representative datasets and employ techniques to mitigate algorithmic bias. For example, an AI designed to diagnose skin conditions should be trained on a wide spectrum of skin tones to ensure accurate performance across all patient populations. Regular audits of AI performance across different demographic groups are essential to identify and address any emergent inequities.

Patient safety is another critical ethical consideration. While AI has the potential to reduce diagnostic errors, it also introduces new risks. An AI that makes a false positive diagnosis could lead to unnecessary anxiety, further testing, and potentially invasive procedures for a patient. Conversely, a false negative could result in a missed diagnosis and delayed treatment, with potentially severe consequences. A human-centered approach prioritizes minimizing these risks through robust validation and continuous monitoring of AI performance in real-world clinical settings. This involves extensive clinical trials, post-market surveillance, and a clear framework for reporting and addressing adverse events related to AI use. The focus is on a gradual, evidence-based deployment of AI, ensuring that patient well-being remains the ultimate priority.

Furthermore, the communication of AI-driven insights to patients must be handled with care and transparency. While the clinician remains the primary point of contact, understanding how AI has informed a diagnosis can be important for patient trust and engagement. Ideally, the AI's role in the diagnostic process should be communicable in a way that is understandable to patients, without overcomplicating or alarming them. This might involve explaining that advanced computer analysis was used to assist the doctor in identifying subtle signs, or that AI helped to assess the probability of certain conditions based on a comprehensive review of their medical information. The goal is to foster confidence in the diagnostic process by demonstrating that it leverages the most advanced tools available, while always being guided by human expertise and empathy.

The development of human-centered AI in medical diagnostics is an ongoing process, requiring collaboration between AI developers, clinicians, ethicists, and patients. It is a journey that emphasizes the creation of tools that are not only technically sophisticated but also ethically sound, clinically relevant, and ultimately, beneficial to human health. The future of medical diagnostics will undoubtedly involve a powerful synergy between human intelligence and artificial intelligence, where AI acts as an indispensable partner, empowering clinicians to deliver the highest standard of care to every patient. This collaborative paradigm, grounded in human-centered principles, promises to unlock new frontiers in disease detection, treatment, and personalized medicine, heralding an era of more accurate, efficient, and equitable healthcare. The ultimate measure of success will be the tangible improvement in patient outcomes and the enhanced trust and confidence that both clinicians and patients place in the diagnostic process, knowing that it is augmented by intelligent, transparent, and ethically aligned AI systems.

The proliferation of wearable artificial intelligence for health monitoring marks a significant evolution in personal healthcare, shifting the paradigm from reactive treatment to proactive wellness management. These devices, ranging from sophisticated smartwatches and fitness trackers to embedded biosensors and specialized patches, are designed to continuously collect a rich tapestry of physiological data. Their core promise lies in their ability to provide real-time insights into an individual's health status, empowering them with information previously accessible only through infrequent clinical visits. However, the successful integration of these technologies into daily life hinges on a meticulous human-centered design process that addresses a multifaceted set of challenges and opportunities.

At the forefront of designing wearable AI for health monitoring is the fundamental consideration of user comfort and wearability. A device, no matter how technologically advanced, will fail to achieve widespread adoption if it is cumbersome, irritating, or aesthetically unappealing. This necessitates a deep understanding of human ergonomics and material science. For instance, a continuous glucose monitor (CGM) that must be worn for extended periods needs to be constructed from biocompatible materials that minimize skin irritation and allergic reactions. The adhesive used to secure the sensor must be robust enough to withstand daily activities, including showering and exercise, yet gentle enough to be removed without causing pain or damaging the skin. Similarly, smartwatches and fitness bands must be lightweight, with flexible straps that conform to the natural contours of the wrist, allowing for a comfortable fit throughout the day and night. The size and shape of the device also play a crucial role; bulky or obtrusive designs can be perceived as socially stigmatizing and physically inconvenient, particularly during sleep or while wearing certain types of clothing. Designers must therefore engage in iterative prototyping and user testing, observing how individuals interact with the devices in their natural environments and gathering feedback on tactile sensations, fit, and overall wearability. This goes beyond mere aesthetics; it is about ensuring that the technology becomes a seamless extension of the user, rather than an alien imposition.

Beyond physical comfort, the imperative of data privacy and security forms another critical pillar in the design of wearable AI for health monitoring. These devices collect highly personal and sensitive physiological data, including heart rate, sleep patterns, activity levels, and in some cases, more intimate metrics like blood oxygen saturation or even electrocardiogram (ECG) readings. The trust that users place in these systems is directly proportional to the perceived security of their data. Therefore, robust encryption protocols are essential at every stage of data transmission and storage, from the sensor to the user's smartphone and any cloud-based platforms. Data anonymization and aggregation techniques should be employed wherever possible, particularly when data is used for research or to train AI models, ensuring that individual identities are protected. Clear and transparent privacy policies are paramount, outlining precisely what data is collected, how it is used, who it is shared with, and how users can control their data. The user interface for managing privacy settings should be intuitive and easily accessible, allowing individuals to make informed decisions about their data sharing preferences. Furthermore, developers must be acutely aware of evolving data protection regulations, such as GDPR and HIPAA, and build systems that are compliant from the ground up. A proactive approach to cybersecurity, including regular security audits and vulnerability assessments, is crucial to mitigate the risk of data breaches and maintain user confidence. The design must consider the entire data lifecycle, from collection and processing to storage and eventual deletion, ensuring that privacy is not an afterthought but a foundational principle.

The accurate interpretation of biometric signals is the

technological bedrock upon which the utility of wearable AI rests. Sensors themselves are only the first step; the true challenge lies in transforming raw data into meaningful health insights. Physiological signals are often noisy, subject to artifacts from movement, environmental factors, and individual physiological variability. For example, a sudden spike in heart rate recorded by a wrist-worn device could be due to an intense workout, a moment of stress, or simply the sensor's inability to accurately track the pulse during rapid arm movement. AI algorithms are indispensable here, acting as sophisticated interpreters of this complex data. Machine learning models, trained on vast and diverse datasets, can learn to distinguish between genuine physiological events and measurement noise. This involves developing algorithms that can:

Denoise and Filter Signals: Employing signal processing techniques to remove unwanted artifacts and isolate the true physiological signal. This might involve Kalman filters for tracking heart rate or specialized algorithms for identifying and discarding motion-induced inaccuracies in ECG readings.
Contextualize Data: Understanding that a particular biometric reading is only meaningful within its context. A high heart rate during exercise is normal, while the same reading at rest might indicate an underlying issue. AI needs to integrate data from multiple sensors (e.g., heart rate, accelerometer, GPS) and potentially from the user's self-reported activities to provide accurate interpretations.

Identify Patterns and Trends: Moving beyond single data points to recognize longer-term patterns and deviations from an individual's baseline. For instance, a gradual increase in resting heart rate over several weeks, even if within the "normal" range, could be an early indicator of developing cardiovascular strain.
Personalize Baselines: Recognizing that "normal" varies significantly from person to person. An AI should establish a personalized baseline for each user based on their historical data and gradually adapt as their physiology changes. This is crucial for detecting subtle, yet significant, individual deviations.
Validate and Calibrate: Ensuring that the data collected by wearable sensors aligns with more established medical measurement devices. Periodic calibration and validation against clinical-grade equipment are necessary to maintain accuracy and build confidence in the device's readings.

This accurate interpretation is vital for generating reliable insights. For example, an AI that misinterprets a brief cardiac arrhythmia as a normal variation could have serious consequences, leading to a missed diagnosis of a potentially life-threatening condition. Conversely, falsely flagging benign fluctuations as problematic can lead to unnecessary user anxiety and a devaluation of the device's alerts. Therefore, the development of these interpretation algorithms must involve close collaboration between AI engineers, biomedical engineers, and clinicians to ensure scientific rigor and clinical relevance.

The ultimate goal of wearable AI for health monitoring is to provide actionable insights to both users and healthcare providers. Raw data, even if accurately interpreted, is of limited value if it doesn't inform decisions or inspire behavior change. This requires translating complex biometric information into clear, understandable, and, most importantly, actionable recommendations.

For the
end-user, these insights should:

Empower Proactive Health Management: Instead of simply displaying a heart rate, the AI might suggest, "Your resting heart rate has been elevated for the past three days. Consider reducing your caffeine intake and ensuring you get adequate sleep to support your cardiovascular health." Or, "Your sleep quality score has decreased significantly. Try to maintain a consistent bedtime and create a relaxing pre-sleep routine."
Provide Timely Alerts: Wearable AI can serve as an early warning system. A device detecting a significant drop in blood oxygen saturation might alert the user to seek medical attention, or a watch that monitors for falls could automatically notify emergency contacts if a sudden, hard impact is detected. For individuals with chronic conditions like atrial fibrillation, an ECG-capable wearable can alert them to a potential episode, prompting them to consult their doctor.

Offer Personalized Health Recommendations: Based on the user's unique data profile, the AI can offer tailored advice. For example, if a user consistently struggles with sleep, the AI could suggest specific sleep hygiene practices that have proven effective for individuals with similar sleep patterns. If activity levels are consistently low, the AI might recommend specific types of exercises that are suitable for their fitness level and preferences, perhaps even integrating with popular fitness apps to suggest guided workouts.
Motivate Behavior Change: Gamification elements, personalized goal setting, and positive reinforcement can be incorporated into the AI's feedback loop to encourage users to adopt healthier habits. Progress tracking and visual representations of achievements can foster a sense of accomplishment and sustained engagement.
Facilitate Self-Advocacy: By providing users with objective data about their health, wearables empower them to have more informed conversations with their healthcare providers. They can present their data logs, discuss observed trends, and actively participate in their care decisions.

For
healthcare providers, wearable AI data can offer a more comprehensive and continuous view of a patient's health beyond the limited snapshots captured during office visits. This data can:

Inform Clinical Decision-Making: Trends in heart rate variability, activity levels, or sleep quality can provide valuable context for diagnosing and managing chronic conditions like hypertension, diabetes, or heart failure. For instance, a physician monitoring a patient with heart failure might use wearable data to track fluid retention (indicated by changes in activity or sleep) and adjust medication proactively.

Identify At-Risk Individuals: AI algorithms can analyze aggregated data from many patients to identify individuals who may be at higher risk for developing certain conditions or experiencing adverse events, allowing for early intervention.

Monitor Treatment Efficacy: Physicians can track how patients respond to treatments in real-time. If a new medication for blood pressure is prescribed, wearable data can show its actual impact on the patient's daily blood pressure readings and heart rate, allowing for timely adjustments.

Support Remote Patient Monitoring (RPM): Wearable devices are a cornerstone of RPM programs, enabling healthcare systems to remotely monitor patients with chronic illnesses. This can reduce hospital readmissions, improve patient outcomes, and lower healthcare costs. The AI component is crucial for sifting through the vast amount of data generated by RPM patients to flag those requiring immediate attention.

Facilitate Research: Anonymized and aggregated data from wearable devices can be a rich source for epidemiological studies and clinical research, helping to identify new health insights and develop more effective treatments.

The integration of these actionable insights requires careful consideration of the user interface and user experience (UI/UX). The information presented must be intuitive, avoiding overly technical jargon. Visualizations, such as graphs and charts that clearly depict trends and deviations, are far more effective than raw numbers. Alerts should be designed to be informative but not alarming, providing context and suggesting appropriate actions. The ability for users to easily share their data with their healthcare providers, perhaps through secure portals or standardized data formats, is also crucial for seamless integration into the broader healthcare ecosystem.

The human-centered design philosophy extends to ensuring that wearable AI devices are inclusive and accessible to a diverse range of users. This means considering individuals with different physical abilities, technological literacy levels, and cultural backgrounds. For example:

Accessibility for Older Adults: Devices should have larger fonts, clear button interfaces, and voice command capabilities. The setup process should be simple and may require assistance from family members or caregivers. The focus might be on fall detection, medication reminders, and basic vital sign monitoring.
Inclusivity for Diverse Populations: As mentioned earlier, AI algorithms must be trained on diverse datasets to ensure accurate performance across all skin tones, ages, and genders. The design of the device itself should consider different body types and needs. For

example, a device designed for a wider wrist may not fit comfortably on a smaller wrist.

Cultural Sensitivity: Health beliefs and practices vary across cultures. The way health information is presented and the types of recommendations made should be sensitive to these differences. For instance, dietary advice needs to consider local food availability and cultural dietary norms.

Addressing Digital Literacy Gaps: Not everyone is comfortable with smartphones or complex apps. The wearable AI system should offer tiered levels of functionality and support, with simpler interfaces for less tech-savvy users and more advanced features for those who desire them. This might include basic functions accessible directly on the device, with optional app-based features for deeper engagement.

Furthermore, the design must anticipate and address potential ethical dilemmas that arise from the continuous collection of personal health data. Beyond privacy, this includes issues of:

Over-reliance and Misinterpretation: Users might place undue trust in the device's readings, leading to self-diagnosis or ignoring professional medical advice. Designers must incorporate disclaimers and emphasize that the wearable is a tool to assist health management, not replace a healthcare professional. Providing clear guidance on when to consult a doctor is essential.

Health Anxiety and Obsession: The constant stream of data can, for some individuals, lead to increased anxiety about their health. The AI's feedback mechanisms should be designed to promote well-being and reassurance, rather than exacerbating worries. Gentle nudges towards healthy behaviors are more effective than constant alerts about minor fluctuations.

Data Security and Malicious Use: As discussed, robust security is paramount. The potential for data to be used for discriminatory purposes (e.g., by insurance companies) or to be targeted by cybercriminals is a significant concern. Transparency about data usage and strong legal protections are crucial.

Equity and Access: Wearable AI devices can be expensive, potentially widening the gap in healthcare access between those who can afford them and those who cannot. Designers and manufacturers should consider tiered pricing models, partnerships with healthcare providers for subsidized access, and the development of more affordable yet effective alternatives. The societal benefit of widespread health monitoring is diminished if it only benefits a privileged segment of the population.

The development of AI-powered wearable health monitoring devices is not a static process but an evolving journey. Continuous research and development are needed to improve sensor accuracy, enhance AI interpretation capabilities, and explore new frontiers in biometric sensing. This includes developing non-invasive or minimally invasive sensors for a wider range of biomarkers, such as stress hormones, inflammation markers, or even early cancer indicators. The integration of advanced AI techniques like federated learning can allow models to be trained across multiple devices without centralizing sensitive user data, further enhancing privacy.

Ultimately, the success of wearable AI in health monitoring will be measured not just by technological sophistication but by its ability to positively impact human lives. A human-centered approach ensures that these powerful tools are designed to be safe, effective, private, equitable, and empowering, fostering a future where individuals are more informed and proactive participants in their own health journeys, supported by intelligent technology that seamlessly integrates into their daily lives. The vision is one where these devices become trusted companions, providing continuous, personalized guidance that contributes to longer, healthier, and more fulfilling lives.

The integration of Artificial Intelligence (AI) into healthcare promises a revolution in diagnostics, treatment, and patient care, offering unprecedented levels of precision and efficiency. However, this transformative potential is shadowed by a complex landscape of ethical considerations that demand careful navigation. As AI systems become increasingly embedded in clinical workflows and patient interactions, it is imperative to critically examine the challenges they present to ensure that technological advancement serves humanity's best interests, particularly in the sensitive realm of health.

At the forefront of these ethical concerns is the safeguarding of patient data privacy and security. AI algorithms, by their very nature, thrive on vast datasets. In healthcare, these datasets are deeply personal, encompassing intimate details about an individual's physical and mental well-being, genetic predispositions, lifestyle habits, and medical history. The collection, storage, and processing of such sensitive information raise significant questions. While robust anonymization and encryption techniques are foundational, the sheer volume and interconnectedness of health data within AI systems create new vulnerabilities. A data breach in a healthcare context is not merely a financial risk; it can lead to identity theft, discrimination, reputational damage, and profound psychological distress for affected individuals. Furthermore, the sophisticated nature of AI means that even anonymized data can sometimes be de-anonymized or linked to other datasets to re-identify individuals, a risk that grows with the increasing availability of public and private data repositories. Consent mechanisms, therefore, must evolve beyond standard checkboxes. True informed consent requires clear, understandable explanations of how patient data will be used, who will have access to it, the potential risks involved, and the duration for which the data will be retained. Patients must have granular control over their data, with the ability to revoke consent and request data deletion, and the technical infrastructure must support these rights without compromising system integrity. The "black box" nature of some AI models also poses a challenge; if even the developers cannot fully explain how a decision is reached, it becomes difficult to assure patients about the ultimate disposition of their information. Establishing clear data governance frameworks, adhering to stringent regulatory requirements like HIPAA and GDPR, and fostering a culture of data stewardship within healthcare institutions are not optional but essential prerequisites for the ethical deployment of AI in this sector. The continuous evolution of cyber threats necessitates perpetual vigilance, with regular security audits, penetration testing, and proactive threat intelligence becoming standard operational procedures rather than reactive measures.

Beyond privacy, the pervasive issue of algorithmic bias poses a critical threat to equitable healthcare. AI models learn from the data they are trained on. If this data reflects historical societal inequities, existing biases will be amplified and perpetuated by the AI system, potentially leading to discriminatory outcomes. For instance, if a diagnostic AI is trained primarily on data from a specific demographic group, it may perform poorly when applied to individuals from underrepresented populations. This could manifest as delayed or missed diagnoses, inappropriate treatment recommendations, or unequal access to life-saving interventions. Examples are already emerging: studies have shown that some algorithms used to predict healthcare needs, when biased towards healthier populations (often implicitly white and male), disproportionately allocate fewer resources to Black patients who, due to existing health disparities, may present with more severe conditions for the same predicted score. Similarly, AI systems trained on biased image datasets might struggle to accurately identify skin conditions on darker skin tones, or facial recognition systems used for patient identification could exhibit higher error rates for certain ethnic groups. Addressing algorithmic bias requires a multi-pronged approach. Firstly, it demands meticulous attention to data collection and curation, ensuring that training datasets are diverse, representative, and free from historical biases wherever possible. This involves actively seeking out data from a wide range of demographics, geographic locations, and socioeconomic backgrounds. Secondly, it requires the development and implementation of bias detection and mitigation techniques during the AI model development and validation phases. This could include fairness-aware machine learning algorithms, differential privacy techniques, and rigorous testing across various subgroups. Regular audits of deployed AI systems are also crucial to monitor for emergent biases in real-world performance. Transparency about the limitations of AI systems, including their potential for bias, is also paramount. Healthcare providers must be educated on these limitations and empowered to critically evaluate AI outputs, rather than accepting them as infallible pronouncements. The goal is not to achieve perfect neutrality, which may be an unattainable ideal, but to strive for fairness and equity, ensuring that AI enhances, rather than erodes, the principles of justice in healthcare.

The question of accountability for AI errors is another complex ethical quandum. When an AI system makes a mistake – misdiagnoses a condition, recommends an incorrect treatment, or fails to flag a critical alert – who is responsible? Is it the AI developer, the institution that deployed the AI, the clinician who relied on its output, or the AI itself? The distributed nature of AI development and deployment, coupled with the "black box" problem, can make pinpointing responsibility exceptionally difficult. Traditional legal and ethical frameworks, designed for human error, often struggle to accommodate the unique characteristics of AI failures. For example, if an AI algorithm is trained on flawed data and subsequently makes a harmful recommendation, is the fault with the data provider, the data scientist who selected the data, the engineer who wrote the algorithm, or the physician who trusted the algorithm's output without independent verification? Establishing clear lines of accountability is vital for fostering trust and ensuring that recourse is available to patients who suffer harm. This may necessitate the development of new legal precedents and regulatory frameworks specifically tailored to AI in healthcare. Some proposals suggest a shared responsibility model, where developers, healthcare providers, and potentially even regulators all bear a degree of accountability. Furthermore, robust oversight mechanisms are needed. This includes establishing independent bodies to review AI systems before their deployment, mandating post-deployment monitoring and incident reporting, and ensuring that AI systems are designed with built-in safeguards and audit trails that can help reconstruct the decision-making process leading to an error. Clinicians must be trained not only on how to use AI tools but also on their limitations and the importance of maintaining their professional judgment. The ultimate decision regarding patient care must remain with a human clinician, who can integrate AI-generated insights with their own expertise, patient context, and ethical considerations. This symbiotic relationship, where AI serves as an intelligent assistant rather than an autonomous decision-maker, is key to mitigating the risks associated with AI errors.

The introduction of AI into the patient-physician relationship also presents a profound ethical challenge. The traditional relationship is built on trust, empathy, and nuanced communication. AI, while capable of processing vast amounts of information and identifying patterns invisible to the human eye, lacks the capacity for genuine human connection. There is a concern that an over-reliance on AI could depersonalize healthcare, reducing patients to sets of data points and eroding the humanistic aspects of care. When a physician relies heavily on AI-generated diagnostic summaries or treatment plans, there's a risk that the empathetic dialogue – the careful listening, the non-verbal cues, the shared decision-making process that forms the bedrock of good medical practice – could be diminished. Patients might feel less heard or understood if their interaction is mediated by algorithms. Conversely, AI could also potentially enhance the patient-physician relationship. By automating routine tasks, AI can free up clinicians' time, allowing them to focus more on direct patient interaction and complex problem-solving. AI-powered tools that provide clear, evidence-based summaries of a patient's condition can facilitate more efficient and informed consultations. The key lies in how AI is integrated. If AI is used as a tool to augment the physician's capabilities, providing them with better information and insights to inform their decisions and communication, it can be beneficial. However, if AI is seen as a replacement for physician expertise or judgment, or if its outputs are presented to patients without proper context and explanation, it could indeed damage the relationship. Open communication about the role of AI in diagnosis and treatment is essential. Physicians should explain to patients when AI is being used, what its purpose is, and how it is informing their recommendations. This transparency helps manage patient expectations and maintains trust. The development of "explainable AI" (XAI) is particularly relevant here, aiming to make AI decisions more transparent and understandable to both clinicians and patients, thereby preserving the human element in care.

Mitigating these ethical risks requires a proactive and collaborative approach. Robust consent mechanisms are not just a legal requirement but an ethical imperative. They must be dynamic, transparent, and provide users with genuine agency over their data. Regular audits for algorithmic bias are essential, not as a one-time check, but as an ongoing process throughout the AI system's lifecycle. These audits should involve diverse teams of experts and be independently verified. Clear guidelines for the use of AI in clinical settings are needed, defining the scope of its application, the level of human oversight required, and the procedures for addressing errors and adverse events. These guidelines should be developed through broad consultation involving AI developers, healthcare professionals, ethicists, policymakers, and patient advocacy groups. Furthermore, fostering a culture of responsible innovation is paramount. This means prioritizing ethical considerations from the initial design phase through deployment and ongoing maintenance. It involves continuous education and training for healthcare professionals on AI capabilities, limitations, and ethical implications. It also means investing in research to better understand the societal impact of AI in healthcare and developing best practices to ensure its equitable and beneficial application. Ultimately, the ethical integration of AI into healthcare hinges on a commitment to human-centered design, where technology serves to augment human capabilities and uphold the core values of medicine: beneficence, non-maleficence, autonomy, and justice. The promise of AI in healthcare is immense, but its realization depends on our collective ability to navigate its ethical complexities with wisdom, foresight, and an unwavering dedication to patient well-being.

The journey towards truly beneficial and widely adopted AI in healthcare necessitates a fundamental shift in its developmental paradigm. Historically, technological advancements, including those in medical devices, have often followed a linear path where engineers and developers create solutions based on their perceived needs and technical capabilities, with end-users engaged only at the later stages for feedback and testing. However, the increasing complexity and intimate nature of AI applications in healthcare demand a more inclusive and collaborative approach. This subsection delves into the critical importance of proactively involving a diverse range of stakeholders – most importantly, patients, but also clinicians and caregivers – throughout the entire lifecycle of healthcare AI development. This isn't merely about gathering feedback; it's about embedding their experiences, values, and practical realities into the very architecture of the AI systems we create. The overarching goal is to move beyond developing AI *for* people to developing AI *with* people, ensuring that these powerful tools are not only clinically sound but also deeply resonant with the lived experiences of those they are intended to serve.

The concept of co-design, also known as participatory design, stands as a cornerstone for achieving this inclusive development model. Co-design actively brings together developers, clinicians, patients, and caregivers in a series of workshops and iterative design sessions. In these settings, participants are not passive recipients of information but active contributors, sharing their insights, challenges, and aspirations related to healthcare. For instance, imagine a workshop focused on an AI-powered diagnostic tool for a chronic condition. Clinicians might articulate the need for seamless integration into existing electronic health records and the importance of a clear, actionable summary of findings. Simultaneously, patients living with the condition could highlight concerns about the AI's output being too technical, the lack of emotional support in its messaging, or the potential for it to increase their anxiety if results are presented without adequate context or human reassurance. Caregivers, often the backbone of support for patients, might bring forth perspectives on managing medication adherence assisted by AI, or how AI could streamline communication with healthcare providers when the patient is unable to do so themselves. These sessions allow for the identification of potential usability barriers, ethical dilemmas, and unmet needs that might otherwise be overlooked. The iterative nature of co-design means that prototypes and concepts are developed, reviewed, and refined based on this continuous input, fostering a shared sense of ownership and ensuring that the final product is a far more robust and relevant solution than one conceived in isolation.

To institutionalize patient voices in the ongoing development process, the establishment of patient advisory boards is an invaluable strategy. These boards comprise individuals who represent various patient demographics, conditions, and healthcare experiences. They serve as ongoing conduits for feedback, providing guidance on the strategic direction of AI product development, scrutinizing user interfaces, and evaluating the ethical implications of proposed features. For an AI system designed to predict and manage hospital readmission risks, a patient advisory board could offer critical insights. They might advise on how the AI's predictive scores are communicated to patients – advocating for a gentle, empowering tone rather than one that instills fear or blame. They could also suggest features that facilitate patient self-management, such as personalized reminders for follow-up appointments or educational content tailored to their specific risks, all framed within a supportive and understandable context. Furthermore, these boards can act as early warning systems for potential biases. For example, if an AI algorithm for predicting disease progression is being developed, a diverse board could flag concerns if certain patient groups feel the AI's predictions are consistently more negative or less optimistic for them, prompting developers to investigate underlying data biases. These advisory roles ensure that the patient perspective is not a singular event but a continuous presence, guiding the evolution of healthcare AI with a consistent focus on patient well-being and empowerment.

Usability testing is a critical phase in ensuring that AI healthcare solutions are not only functional but also accessible and easy to use for their intended audience. This phase becomes exponentially more impactful when conducted with diverse patient groups, encompassing individuals with varying levels of digital literacy, different age groups, those with cognitive or physical disabilities, and individuals from different cultural and linguistic backgrounds. An AI-powered remote monitoring system for elderly patients with heart conditions, for instance, requires meticulous usability testing. A young, tech-savvy patient might navigate the application with ease, but an 80-year-old with visual impairments and limited experience with smartphones might struggle with small font sizes, complex navigation menus, or unfamiliar icons. Testing with such a diverse group allows developers to identify specific points of friction. This could lead to design modifications like larger buttons, voice-guided instructions, simplified workflows, and clear, jargon-free language. It also highlights the need for accessible data presentation; AI-generated summaries of physiological data should be easily interpretable by both the patient and their caregiver, perhaps with visual aids and clear indications of what requires immediate attention from a healthcare professional. Beyond interface design, usability testing can reveal how patients actually interact with AI outputs. Do they trust the AI's recommendations? Do they feel empowered to ask questions about the AI's findings? Or do they feel overwhelmed and disempowered by the technology? The answers to these questions are vital for shaping the AI's communication style, the accompanying educational materials, and the overall user experience to foster confidence and encourage adherence to treatment plans.

The integration of these participatory methods is not merely about good practice; it is intrinsically linked to building trust and enhancing the adoption of AI technologies in healthcare. When patients and clinicians are involved in the design process, they develop a deeper understanding of how the AI works, its capabilities, and its limitations. This transparency demystifies the technology and reduces apprehension, fostering a sense of partnership rather than a perceived imposition of technology. For AI systems that influence diagnostic pathways or treatment recommendations, this trust is paramount. A clinician who has been involved in the AI's development, perhaps by contributing clinical expertise or participating in validation studies, is more likely to integrate the AI's insights into their practice with confidence, understanding its strengths and knowing when to critically evaluate its outputs. Similarly, patients who have had their concerns addressed during the design phase, and whose needs have been demonstrably considered, are more likely to accept and utilize AI-driven health management tools. For example, an AI chatbot designed to provide mental health support might gain greater acceptance if individuals with lived experience of mental health challenges have contributed to its conversational design, ensuring its tone is empathetic, non-judgmental, and culturally sensitive, and that it clearly outlines when to seek professional human help. This active involvement cultivates a sense of shared responsibility and alignment with the AI's purpose, moving beyond a purely transactional relationship with technology to one built on mutual understanding and shared goals.

Moreover, the practical application of these inclusive methodologies directly translates into improved health outcomes. AI systems that are designed with a deep understanding of user needs and preferences are inherently more likely to be used correctly and consistently. This leads to more accurate data collection, better adherence to treatment recommendations, and more effective management of chronic conditions. Consider an AI-driven medication management system. If it's designed in consultation with patients and pharmacists, it can incorporate features that address common adherence challenges, such as personalized reminders that account for patient routines, clear instructions on how to take medication, and alerts for potential drug interactions that are easily understood. Such a system, shaped by real-world user input, is far more likely to achieve its intended effect – improving medication adherence – than a system developed solely from a technical standpoint. Similarly, AI tools for chronic disease management, developed with input from patients and their care teams, can be tailored to provide support that is both clinically relevant and personally motivating, leading to better self-management, fewer exacerbations, and improved quality of life. The emphasis on building AI *with* people, rather than *for* them, ensures that the technology becomes an integrated, supportive element of care, rather than an intrusive or alienating force. It fosters an environment where AI amplifies human capabilities and empathy, ultimately serving the fundamental goal of improving health and well-being for all. This human-centered approach, where the user is a co-creator, is not just an ethical ideal but a pragmatic necessity for unlocking the full transformative potential of AI in healthcare.

The practical implementation of these participatory approaches requires a conscious effort to overcome potential barriers. Developers might initially find the inclusion of diverse stakeholders to be time-consuming and resource-intensive. However, the long-term benefits of avoiding costly redesigns, improving user adoption rates, and ultimately achieving better health outcomes far outweigh these initial investments. Clear communication protocols must be established to ensure that feedback from various groups is effectively captured, understood, and integrated into the development process. This often involves employing skilled facilitators who can bridge the gaps between technical jargon and everyday language, and ensuring that all participants feel heard and valued. Furthermore, it is crucial to acknowledge that not all stakeholders have the same level of technical understanding or the same ability to articulate their needs in a way that is immediately translatable into design specifications. Therefore, the process needs to be flexible, allowing for different forms of input – from spoken narratives and sketches to direct observation of user behavior. The goal is to translate lived experiences into actionable design requirements, a task that requires empathy, patience, and a genuine commitment to understanding the user's world.

Moreover, the ethical implications of not involving end-users are significant. Developing AI systems in a vacuum can inadvertently lead to solutions that are biased, inaccessible, or even harmful. Without the direct input of diverse patient populations, algorithms may perpetuate existing health disparities, fail to account for cultural nuances in health beliefs or practices, or be designed in ways that exclude individuals with specific disabilities. For instance, an AI symptom checker that uses a purely text-based interface might be inaccessible to individuals with visual impairments, or an AI that provides diagnostic probabilities without clear explanations might confuse or distress patients from cultures where direct communication of potentially negative news is handled with extreme caution. The principle of "nothing about us without us" is particularly pertinent here. It underscores that decisions about health technologies, especially those that deeply impact individual well-being, should not be made without the direct and meaningful participation of the people affected by those decisions. This participatory ethos is not merely about optimizing usability; it is a fundamental aspect of ensuring equity, justice, and respect for autonomy in the design and deployment of healthcare AI. By actively involving patients, clinicians, and caregivers as co-creators, we lay the groundwork for AI solutions that are not only technically sophisticated but also ethically sound, socially responsible, and ultimately, more effective in improving the health and lives of all. This collaborative construction of AI in healthcare fosters a system where technology serves humanity, guided by the collective wisdom and lived experiences of those it aims to serve.

The integration of artificial intelligence into healthcare is rapidly transforming patient care, moving beyond diagnostic tools and treatment planning to encompass a wide array of therapeutic interventions. Among the most promising applications are AI-assisted devices designed to enhance and personalize physical rehabilitation. These technologies are not just augmenting traditional therapy sessions; they are redefining the landscape of recovery by offering more tailored, engaging, and data-driven approaches to help individuals regain mobility and independence. This subsection delves into a specific case study to illuminate the profound impact of human-centered AI in the realm of physical rehabilitation.

One significant area where AI is making substantial inroads is in the development of intelligent rehabilitation devices for individuals recovering from injuries, surgeries, or chronic conditions affecting their motor functions. Consider the challenges faced by a patient undergoing physical therapy after a stroke. Traditional therapy often involves repetitive exercises performed under the guidance of a therapist. While effective, this approach can be limited by the therapist's time, the patient's endurance, and the subjective nature of progress assessment. AI-powered devices aim to address these limitations by providing a more dynamic, personalized, and continuously monitored therapeutic experience.

A prime example of such innovation is an AI-driven system designed for upper limb rehabilitation. This system typically comprises a wearable sensor array, often integrated into gloves or armbands, and a sophisticated software platform that analyzes the sensor data. The sensors capture a rich stream of information, including joint angles, muscle activation patterns (electromyography or EMG), movement speed, and range of motion. The AI engine then processes this data in real-time, comparing it against a personalized baseline established during initial assessments and against established biomechanical models.

The personalization aspect is where AI truly shines. Instead of a one-size-fits-all exercise regimen, the AI can dynamically adjust the difficulty and type of exercises based on the patient's performance. For instance, if a patient is struggling to achieve a certain range of motion, the AI can simplify the task, provide more explicit visual or haptic cues, or even break down the movement into smaller, more manageable steps. Conversely, if the patient is performing well and demonstrating improved strength or coordination, the AI can gradually increase the challenge to promote further recovery and prevent plateaus. This adaptive nature ensures that the therapy remains optimally challenging, maximizing engagement and accelerating progress.

Motivation is a critical factor in rehabilitation, as recovery can be a long and arduous process. Many patients find traditional exercises monotonous, leading to decreased adherence and slower progress. AI-assisted devices combat this by incorporating gamification elements and providing engaging feedback. The software platform can transform exercises into interactive games, where successful completion of therapeutic movements leads to rewards within the game, such as advancing to a new level, scoring points, or unlocking virtual achievements. The real-time visual feedback on screen, showing the patient's performance relative to the task or game objective, further enhances engagement. The AI can also provide immediate, encouraging messages, celebrating small victories and reinforcing positive effort, which can be incredibly motivating for individuals facing significant physical challenges.

Beyond exercise and motivation, accurate progress tracking is paramount for both patients and therapists. AI systems can provide objective and granular data that surpasses traditional therapist observations. The sensors continuously collect data, creating a detailed log of every movement, every session. The AI analyzes this data to quantify improvements in strength, coordination, speed, and endurance over time. This objective data allows therapists to gain a deeper understanding of a patient's progress, identify specific areas of weakness or improvement, and make more informed decisions about the therapy plan. For the patient, seeing tangible, data-driven evidence of their recovery can be a powerful motivator and provide a clear picture of their journey. The AI can generate comprehensive reports, often presented in easily understandable visual formats, highlighting key metrics and trends.

The feedback loop extends to the therapist as well. The AI platform can alert therapists to potential issues, such as abnormal movement patterns that might indicate compensatory strategies or increased risk of re-injury. It can also provide insights into the patient's engagement levels and adherence outside of supervised sessions. This allows therapists to provide more targeted interventions and support. For example, if the AI detects a consistent issue with a particular muscle group during home-based exercises, the therapist can focus on addressing that specific deficit during the next in-clinic session. This collaborative intelligence between the AI and the human therapist ensures a more holistic and effective approach to rehabilitation.

The design process for these AI-assisted rehabilitation devices is inherently user-centered, with a strong emphasis on the needs and experiences of patients and therapists. Early in the development, extensive user research is conducted. This involves observing patients in rehabilitation settings, interviewing individuals recovering from various conditions, and consulting with physical therapists. Key user needs that emerge consistently include the desire for tangible progress, the need for clear and understandable instructions, the importance of feeling supported and motivated, and the necessity of seamless integration into their daily lives.

For patients, ease of use is critical. A complex setup process or an unintuitive interface can become a barrier to adoption. Therefore, the design prioritizes simple donning and doffing of wearable sensors, straightforward software interfaces with clear visual cues, and minimal need for technical expertise. The AI is trained to interpret a wide range of user inputs, even those that are slightly imprecise, reflecting the reality of impaired motor control.

Therapists, on the other hand, require tools that are efficient, provide actionable insights, and integrate well with their existing workflow. They need to trust the accuracy of the AI's data and its recommendations. The AI's ability to summarize complex data into easily digestible reports, flag potential concerns, and suggest adjustments to the therapy plan makes it a valuable assistant rather than a replacement for their expertise. The collaborative nature of the AI, where it acts as an intelligent aid rather than an autonomous decision-maker, fosters greater acceptance and adoption among clinicians.

Consider a different but equally impactful application: AI-powered exoskeletons for gait rehabilitation after spinal cord injury or neurological disorders. These devices provide mechanical assistance to support leg movements, allowing patients to stand and walk. The AI component in these exoskeletons is crucial for adapting to the user's intent and physiological state. By analyzing sensor data from the user's body (e.g., subtle shifts in weight, muscle activity) and the environment (e.g., detecting inclines or stairs), the AI can predict the user's next movement and provide the appropriate level of assistance. This intelligent adaptation is what differentiates an AI-assisted exoskeleton from a purely mechanical one, enabling a more natural and fluid gait pattern, thereby enhancing the effectiveness of the rehabilitation process. The AI learns the user's gait over time, becoming more attuned to their specific needs and capabilities, which is vital for promoting muscle re-education and improving functional recovery.

Furthermore, the AI can monitor the biomechanics of

the user's gait, identifying inefficiencies or abnormal patterns that might be developing. This data can then be used by therapists to fine-tune the exoskeleton's settings or to design targeted exercises to address these issues. The feedback provided by the AI can also guide the patient, helping them to develop a more balanced and symmetrical gait. The ultimate goal is to facilitate the gradual reduction of reliance on the exoskeleton as the patient's own motor control and strength improve.

Another important facet of AI-assisted rehabilitation is the potential for remote monitoring and telehealth. AI-powered devices can collect data from patients in their homes, allowing for continuous monitoring and intervention without the need for frequent in-person visits. This is particularly beneficial for individuals living in rural areas or those with mobility challenges. The AI can analyze the data remotely, identifying any deviations from the expected recovery trajectory and alerting the therapist if intervention is needed. This not only increases accessibility to rehabilitation services but also allows for more proactive and timely care, potentially preventing complications and improving overall outcomes.

The AI can also facilitate virtual reality (VR) based rehabilitation. VR environments can be used to create immersive and engaging therapeutic scenarios that are difficult or impossible to replicate in a traditional therapy setting. AI can personalize these VR experiences based on the patient's progress and needs, creating custom challenges and feedback mechanisms. For example, an AI could dynamically adjust the difficulty of a VR task, such as a virtual grocery shopping simulation, based on the patient's performance in controlling their movements and maintaining balance. This fusion of AI and VR offers a powerful new paradigm for rehabilitation, making therapy more engaging, effective, and adaptable.

The ethical considerations in the design and deployment of these AI-assisted rehabilitation devices are also significant. Transparency about how the AI works, what data it collects, and how it is used is crucial for building trust with both patients and therapists. Ensuring data privacy and security is paramount, especially when dealing with sensitive health information. Furthermore, developers must be vigilant about potential biases in the AI algorithms, ensuring that the devices are effective and equitable for all individuals, regardless of their background or demographic characteristics. The goal is to ensure that AI enhances rehabilitation for everyone, not just a select few.

In conclusion, AI-assisted rehabilitation devices represent a significant leap forward in restoring function and improving the quality of life for individuals recovering from various physical impairments. By leveraging AI for personalization, motivation, objective tracking, and intelligent feedback, these technologies are transforming the therapeutic landscape. The human-centered design approach, which prioritizes user needs and fosters collaboration between AI and human expertise, is key to unlocking the full potential of these innovations. As AI continues to evolve, we can anticipate even more sophisticated and accessible rehabilitation solutions that empower individuals on their journey to recovery and independence. The integration of AI is not merely about augmenting existing therapies but about fundamentally reimagining what is possible in the pursuit of physical well-being.

Chapter 12: Practical Application: Assistive Technologies

Artificial intelligence is emerging as a transformative force in the realm of assistive technologies, opening new avenues for individuals with disabilities to navigate their environments, communicate effectively, and engage more fully with society. This section delves into the profound impact of AI-driven solutions designed to enhance accessibility, focusing on how these technologies are not merely compensating for limitations but actively empowering users and fostering greater independence. The core principle guiding this development is empathy-driven design, a process that deeply considers the unique lived experiences and specific needs of individuals with diverse abilities.

For individuals with visual impairments, the digital world can present significant barriers. Traditional screen readers, while invaluable, often rely on pre-defined interpretations of web page structures or images. AI is revolutionizing this landscape by enabling more nuanced and context-aware assistance. Advanced AI-powered screen readers can now interpret visual information with unprecedented accuracy. This goes beyond simply identifying objects; these systems can describe the content of images, understand the context of a scene, and even infer emotional cues from facial expressions in photos or videos. For example, a visually impaired user browsing online content can receive a detailed, AI-generated description of a photograph accompanying an article, offering insights into its composition, subjects, and overall mood, thereby enriching their understanding and engagement.

Similarly, AI can assist in navigating complex visual interfaces in applications or operating systems that may not be fully optimized for screen reader technology, providing more intuitive and detailed guidance. The ability of AI to perform optical character recognition (OCR) on a broader range of documents, including handwritten notes or complex diagrams, further expands access to information previously out of reach. This intelligent interpretation reduces the cognitive load on the user and provides a richer, more comprehensive experience of digital content.

The development of AI for accessibility is also profoundly impacting individuals with communication impairments. Speech-to-text (STT) and text-to-speech (TTS) technologies, already established assistive tools, are being significantly enhanced by AI. Modern AI-driven STT systems are far more accurate, capable of understanding a wider range of accents, dialects, and speech patterns, even in noisy environments. This means that individuals who have difficulty with clear enunciation, or whose speech is affected by conditions like ALS, cerebral palsy, or stroke, can have their spoken words transcribed with greater fidelity. This improved accuracy translates directly into more effective communication in professional settings, social interactions, and personal correspondence. Furthermore, AI is enabling personalized STT models that can be trained on an individual's unique speech characteristics, leading to near-perfect transcription for that specific user.

Complementing STT, AI-powered TTS systems are generating more natural-sounding and expressive speech. Instead of robotic, monotone voices, AI can now modulate pitch, intonation, and rhythm to convey emotion and emphasis, making synthesized speech more engaging and easier to understand. This is particularly beneficial for individuals who rely on TTS for all their communication needs, as it allows for more nuanced expression and reduces the potential for misinterpretation. Beyond simple vocalization, AI is enabling more sophisticated communication aids. For instance, predictive text and sentence generation, powered by advanced natural language processing (NLP), can anticipate what a user intends to say, significantly speeding up communication for individuals with limited typing or input speed. These systems learn from the user's communication style and common phrases, becoming increasingly efficient over time. Some advanced AI systems are even being developed to translate complex thoughts or emotions into coherent linguistic output, acting as a sophisticated co-pilot in communication.

The field of prosthetics and orthotics is another area where AI is driving significant advancements in assistive technology. Intelligent prosthetics, powered by AI, are moving beyond simple mechanical replacements for limbs to sophisticated devices that can learn and adapt to the user's movements and intentions. AI algorithms analyze data from sensors embedded in the prosthetic, such as muscle signals (EMG), pressure sensors, and motion detectors, to predict the user's desired actions. This allows for more intuitive control and a more natural gait for prosthetic legs, or more dexterous manipulation for prosthetic hands. For example, an AI-powered prosthetic hand can learn to differentiate between grasping a delicate egg and a heavy object, automatically adjusting its grip strength and pressure accordingly. This level of adaptability significantly enhances a user's independence in performing daily tasks, from cooking and dressing to engaging in hobbies. The AI also contributes to the user's proprioception, providing haptic feedback to the user's residual limb, which helps them to better sense the position and movement of the prosthetic, thus improving control and reducing the cognitive effort required.

Beyond these specific examples, AI is being integrated into a wide array of assistive devices and applications to foster greater independence and engagement. For individuals with mobility impairments, AI can enhance the functionality of wheelchairs, making them more responsive to user input, capable of navigating complex environments, or even learning preferred routes. AI can also power smart home devices to a degree that makes them truly accessible. For instance, voice-controlled assistants can manage lighting, temperature, entertainment systems, and security, all controlled through natural language commands, offering a level of autonomy previously unattainable for many. AI can also analyze patterns in a user's environment and behavior to proactively adjust settings or provide reminders, further supporting independent living.

The design philosophy behind these AI-driven assistive technologies is deeply rooted in empathy. Developers and designers engage in extensive user research, co-creation workshops, and iterative testing with individuals with disabilities. This ensures that the technology is not just functional but also user-friendly, aesthetically pleasing, and culturally sensitive. The goal is to create tools that seamlessly integrate into the user's life, becoming natural extensions of their capabilities rather than cumbersome additions. This means considering factors like the weight and comfort of wearable devices, the intuitiveness of user interfaces, the privacy implications of data collection, and the affordability and availability of these solutions. The emphasis is on enabling users to achieve their goals, whether it's participating in the workforce, pursuing education, enjoying social activities, or simply managing daily life with greater ease.

Consider the application of AI in navigation for individuals with visual impairments. Beyond basic GPS, AI can provide detailed, real-time audio descriptions of the surroundings. This includes identifying potential hazards like uneven pavement, overhead obstructions, or busy intersections, and providing context-specific information about nearby businesses or landmarks. AI-powered computer vision can even read signage, bus numbers, and identify the presence and direction of approaching vehicles, offering a level of situational awareness that significantly enhances safety and confidence when traveling independently. The AI can also learn a user's preferred routes and daily routines, offering proactive guidance and suggestions.

In the realm of learning and education, AI is creating more inclusive environments. For students with learning disabilities like dyslexia, AI can provide real-time text simplification, highlight key information, or offer personalized learning pathways that adapt to their pace and learning style. AI-powered tutors can offer patient, non-judgmental support, breaking down complex subjects into manageable steps and providing immediate feedback. For students with hearing impairments, AI can provide highly accurate, real-time captioning for lectures and discussions, ensuring they do not miss crucial information. The ability of AI to analyze student performance data can also alert educators to students who may be struggling, allowing for early intervention and tailored support.

The ethical considerations surrounding AI in assistive technology are paramount. Transparency in how AI systems work, what data they collect, and how that data is used is essential for building trust. Users must have control over their data and understand the potential limitations or biases within AI algorithms. Developers have a responsibility to ensure that these technologies are equitable and do not exacerbate existing societal inequalities. This involves rigorous testing across diverse user groups and proactive efforts to mitigate biases related to race, gender, socioeconomic status, or the specific nature of a disability. The principle of "nothing about us without us" is critical, meaning that individuals with disabilities must be actively involved in every stage of the design, development, and deployment of these technologies.

The future of AI in assistive technology promises even greater integration and personalization. We can anticipate AI systems that are even more anticipatory, understanding user needs before they are explicitly stated. Imagine AI that can predict when a person with a tremor might need a stabilized utensil or when someone with limited mobility might require assistance with a particular task, proactively offering support through connected devices. The convergence of AI with other emerging technologies like augmented reality (AR) also holds immense potential. AR glasses, guided by AI, could overlay real-time information and assistance directly into a user's field of vision, offering interactive navigation, object identification, or even step-by-step instructions for tasks.

Ultimately, the goal is to leverage AI not as a replacement for human connection or support, but as a powerful enabler that removes barriers, enhances capabilities, and fosters a more inclusive and accessible world for everyone. The ongoing advancements in AI are not just about creating smarter devices; they are about creating more equitable opportunities and empowering individuals to live lives of greater independence, dignity, and fulfillment. The focus remains squarely on the human user, ensuring that technology serves to amplify human potential and enrich individual lives.

The evolution of assistive technologies is increasingly focusing on creating intelligent living spaces designed to foster independence and enhance the quality of life for individuals, particularly the elderly and those with mobility impairments. These "smart environments" leverage the power of artificial intelligence to automate daily tasks, enhance safety, and provide proactive support, transforming traditional homes into adaptive and responsive ecosystems. The underlying principle is to create a living space that not only accommodates but anticipates the user's needs, thereby reducing reliance on external assistance and promoting a greater sense of autonomy and dignity.

At the heart of these smart environments are sophisticated home automation systems, often powered by AI, that meticulously manage various aspects of a dwelling. Lighting, for instance, can be dynamically adjusted based on the time of day, ambient light levels, or the user's presence in a room. This not only contributes to energy efficiency but also enhances safety by ensuring adequate illumination, reducing the risk of falls, especially for individuals with visual impairments or reduced mobility. Imagine a system that gradually brightens the bedroom lights in the morning to gently wake the occupant, or automatically illuminates a path to the bathroom during the night.

Temperature control is another crucial element. AI algorithms can learn a user's preferences and adjust heating or cooling systems accordingly, ensuring optimal comfort without requiring manual intervention. This is particularly beneficial for individuals who may find it difficult to operate thermostats or who experience fluctuations in body temperature.

Security and safety alerts form a critical component of these intelligent homes. AI-powered sensors can monitor for a range of potential hazards, from unusual sounds indicating a fall or distress, to detecting smoke or carbon monoxide leaks, or even monitoring for prolonged periods of inactivity that might suggest a medical emergency. These systems can then trigger automated responses, such as alerting emergency services, notifying pre-designated family members or caregivers, or even activating in-home communication devices to check on the occupant. For individuals with cognitive impairments, features like door and window sensors can provide peace of mind by alerting caregivers if a door is opened at an unusual hour or if someone leaves the home unexpectedly. The integration of smart locks can also allow authorized individuals remote access for emergency purposes, further enhancing safety and security.

Designing intuitive interfaces for controlling these complex systems is paramount to their successful adoption. While voice control, powered by natural language processing (NLP), offers a hands-free and accessible method of interaction, it is not always sufficient or appropriate in all situations. Therefore, a multi-modal approach is often employed, incorporating user-friendly touch interfaces on tablets or smartphones, as well as physical buttons or remote controls strategically placed for easy access. The design of these interfaces must prioritize simplicity, clarity, and adaptability. For instance, a user interface might present large, easily recognizable icons for common functions like "lights on/off" or "adjust temperature." Furthermore, these interfaces can be customized to display only the functions that are relevant to the individual user, reducing cognitive load and preventing confusion. The AI can also learn the user's typical commands and preferences, streamlining interactions. For example, if a user consistently asks for the living room lights to be dimmed after 8 PM, the system can be programmed to do so automatically or prompt the user with a simple "Would you like me to dim the lights?" query.

Reliability and security are non-negotiable aspects of smart home technology, especially when lives and well-being are at stake. The systems must be robust, with fail-safe mechanisms to ensure continued operation even in the event of power outages or network disruptions. Data security is equally crucial. Sensitive personal information, including health data and daily routines, is often collected by these systems. Robust encryption protocols, secure authentication methods, and clear data privacy policies are essential to protect users from unauthorized access and potential misuse of their information. Regular software updates are necessary to patch vulnerabilities and maintain system integrity. The AI components themselves must be designed with security in mind, preventing malicious actors from gaining control of the home environment. This might involve localized processing of sensitive data where possible, reducing the reliance on external cloud servers. Rigorous testing under various simulated failure conditions is also a vital part of ensuring the dependability of these systems.

Beyond mere automation and security, the true potential of smart environments lies in their ability to provide proactive assistance. AI algorithms can analyze patterns in a user's behavior and environmental data to anticipate needs and offer support before being explicitly asked. For example, if sensors detect that a user has been inactive for an extended period during their usual waking hours, the system might gently prompt them to get up and move, or offer to initiate a communication with a caregiver. Similarly, the system could learn a user's medication schedule and provide timely reminders, or even integrate with smart pill dispensers. For individuals with cognitive decline, the system could offer step-by-step guidance for daily tasks, such as preparing a meal or following a morning routine, displaying instructions on a nearby screen or through audio prompts. This proactive support can significantly reduce anxiety and enhance a user's ability to manage their daily life independently.

Consider an AI system that monitors a user's gait and movement patterns within the home. By analyzing data from motion sensors or wearable devices, the AI could detect subtle changes that might indicate an increased risk of falls, such as a slower step, a wider base of support, or unsteadiness. The system could then proactively offer interventions, such as suggesting a rest period, reminding the user to use their assistive device, or even subtly adjusting lighting to improve visibility. In the kitchen, an AI could monitor refrigerator contents and suggest meal ideas based on available ingredients and the user's dietary needs, while also reminding them about expiry dates. For individuals with sensory impairments, the system can translate visual information into auditory cues or tactile feedback. For example, a smart doorbell could announce who is at the door through an audio alert, or a smart oven could provide auditory confirmation when a pre-set temperature is reached.

The integration of AI into these environments also extends to social engagement and mental well-being. Smart mirrors could offer personalized greetings and reminders, display news and weather updates, and even facilitate video calls with loved ones. For individuals who experience loneliness or social isolation, AI-powered companions or chatbots can provide conversational interaction and emotional support, acting as a friendly presence in the home. These AI companions can be programmed to engage users in cognitive exercises, share stories, or play music, helping to maintain mental acuity and reduce feelings of isolation. The system can also facilitate connections with community resources or social groups, encouraging active participation in life.

The design of these smart environments requires a deep understanding of human-centered design principles, with a particular focus on the unique needs and preferences of the target user groups. Co-design workshops, where individuals who will use the technology are actively involved in the design process, are crucial for ensuring that the systems are not only functional but also usable, desirable, and respectful of individual autonomy. This iterative process allows for continuous feedback and refinement, ensuring that the technology truly serves its purpose without being intrusive or overwhelming. For instance, users might express preferences for how alerts are delivered – some may prefer gentle chimes, while others might need more insistent audible alarms. The system's ability to adapt to these individual preferences is key to its success.

Furthermore, the ethical implications of such pervasive AI systems must be carefully considered. Transparency regarding data collection and usage is vital. Users should have clear understanding and control over what data is being collected and how it is being used. Consent mechanisms must be robust and easily understandable. Privacy must be protected through strong security measures and data anonymization where appropriate.

There is also a need to guard against over-reliance on technology, ensuring that human connection and support remain central to care. The goal is to augment human capabilities and enhance independence, not to replace human interaction entirely. The AI should be a tool that empowers individuals, allowing them to live fuller, more independent lives, rather than creating a sense of being constantly monitored or controlled. This delicate balance requires ongoing dialogue between technologists, ethicists, users, and caregivers.

The future of smart environments for independent living will likely see even deeper integration of AI with emerging technologies. For example, the fusion of AI with robotics could lead to assistive robots capable of performing more complex physical tasks, such as fetching objects, assisting with personal care, or even providing physical support for mobility. The use of augmented reality (AR) could overlay contextual information and guidance directly into a user's field of vision, offering real-time instructions for tasks or highlighting potential hazards. Wearable biosensors, integrated with AI, could provide continuous health monitoring, detecting early signs of illness or stress and enabling proactive interventions. The development of more sophisticated emotional AI could lead to systems that are better able to understand and respond to a user's emotional state, providing more empathetic and personalized support. As AI continues to advance, the concept of a "smart home" will evolve from a collection of connected devices to a truly intelligent and adaptive living partner, dedicated to promoting health, safety, and well-being, and ultimately enabling individuals to live with greater independence and dignity for as long as possible. This vision necessitates a commitment to ongoing research, ethical development, and user-centric design, ensuring that technology serves humanity's best interests and fosters a more inclusive and supportive society.

The landscape of communication is undergoing a profound transformation, driven by the remarkable advancements in Artificial Intelligence (AI). For individuals who face challenges in expressing themselves verbally or understanding spoken language, AI-powered communication aids are emerging as powerful catalysts for independence, connection, and social inclusion. These technologies are moving far beyond simple text-to-speech or pre-recorded messages, delving into nuanced understanding and generation of communication, thereby bridging significant gaps that previously hindered meaningful interaction.

At the forefront of these innovations is the sophisticated evolution of predictive text and augmented communication systems. Traditional predictive text, while helpful, often relies on generic word lists and basic grammatical structures. Modern AI-driven systems, however, learn from an individual user's unique vocabulary, communication style, and even their emotional context. This deep learning approach allows the AI to anticipate not just the next word, but entire phrases and even sentence structures that the user is likely to employ. Imagine a system that has learned that a particular user, when discussing their gardening hobby, frequently uses specific terms like "pruning," "mulch," and "compost" in a certain sequence. The AI can then offer these predictions with remarkable accuracy, significantly reducing the time and effort required to compose messages. This level of personalization is crucial for individuals who might have limited motor control, making typing a slow and laborious process, or those who struggle with word retrieval due to cognitive conditions. The system becomes an extension of the user's own thought process, offering suggestions that feel intuitive and natural, rather than merely assistive. This learning extends beyond simple word prediction; AI can also infer intent. If a user repeatedly selects a particular phrase in response to certain questions, the AI can learn to offer that phrase proactively when similar contexts arise, further streamlining communication.

Furthermore, AI is revolutionizing the way we interpret and generate non-verbal communication. A particularly groundbreaking area is the application of AI in sign language translation. For the Deaf and hard-of-hearing communities, sign language is a rich and expressive form of communication. However, for individuals who do not know sign language, a significant communication barrier exists. AI is being trained on vast datasets of sign language, encompassing various dialects and individual signing styles. Through advanced computer vision and machine learning algorithms, these systems can now detect hand gestures, facial expressions, and body movements associated with sign language and translate them into spoken words or written text in real-time. This technology holds the potential to enable seamless conversations between sign language users and non-signers in everyday settings, from doctor's appointments and educational lectures to casual social gatherings. The accuracy and speed of these translation systems are continually improving, moving closer to a point where they can facilitate fluid and natural dialogue. The development of sophisticated models that can distinguish subtle nuances in hand shapes, movements, and even the emotional tone conveyed through facial expressions is paramount to achieving truly effective translation. This involves not just recognizing individual signs, but understanding the grammatical structures and idioms inherent to sign languages, which differ significantly from spoken languages.

Conversely, AI is also being employed to translate spoken or written language into sign language. This is achieved through AI-powered avatars or robotic systems that can perform the corresponding signs. This dual capability – translating from sign to speech/text and vice versa – creates a comprehensive communication bridge, fostering greater understanding and inclusion for individuals who rely on sign language. The potential applications are vast, including educational tools that can sign along with spoken lessons, or personal communication devices that can sign a text message received by the user. The ethical considerations in this domain are also significant, ensuring that the AI models are trained on diverse and representative datasets to avoid bias and to accurately reflect the nuances of different sign languages and signing styles. User involvement in the development process is crucial to ensure that these tools are not only technically proficient but also culturally sensitive and respectful of the Deaf community's linguistic heritage.

For individuals with complex speech impairments, often resulting from conditions like cerebral palsy, ALS, stroke, or traumatic brain injury, AI offers a pathway to regain or enhance their ability to express themselves. These individuals may have difficulty with articulation, volume, or the very production of sounds. AI-powered Augmentative and Alternative Communication (AAC) devices are being tailored to their specific needs. These systems can analyze even subtle vocalizations, muscle movements, or eye-gaze patterns and translate them into synthesized speech or text. For instance, a person who can only produce a limited range of sounds might have their vocalizations mapped by an AI to a much broader vocabulary. The AI learns the unique acoustic properties of their speech and identifies patterns that correspond to different words or concepts, enabling them to communicate more complex thoughts.

Eye-gaze technology, coupled with AI, has become increasingly sophisticated. Users can control a communication interface by looking at specific icons or letters on a screen. The AI analyzes their gaze patterns, predicts their intended selection with high accuracy, and can even learn to anticipate what they might want to say based on the context of the conversation or their previous interactions. This allows for remarkably rapid communication for individuals who have no voluntary motor control below the neck. The AI can also help to refine the selection process, distinguishing between intentional selections and accidental glances, thereby reducing frustration and improving the efficiency of communication. Furthermore, some systems are beginning to incorporate emotional AI to interpret the user's intended tone. If the user’s facial expression or subtle vocalizations indicate frustration or happiness, the AI can adjust the synthesized speech output to reflect that emotion, making the communication more natural and nuanced.

Personalized communication interfaces are another area where AI is making a significant impact. Recognizing that no two individuals with communication challenges are exactly alike, AI enables the customization of these interfaces to an unprecedented degree. This goes beyond simply resizing text or changing color schemes. AI can analyze a user's cognitive abilities, motor skills, visual acuity, and personal preferences to create an interface that is optimally suited for them. For example, for someone with limited visual processing, the AI might simplify the interface, use high-contrast colors, and offer larger, more distinct icons. For an individual with memory challenges, the AI might organize frequently used phrases and words in easily accessible locations or provide visual cues to aid recall. The system can learn which communication strategies are most effective for a particular user over time, adapting its suggestions and interface layout accordingly. This dynamic adaptation ensures that the assistive technology remains relevant and effective as the user's needs or abilities change.

The design principles underpinning these AI-powered communication aids are centered on several key pillars: customization, adaptability, and natural interaction. Customization ensures that the technology is tailored to the individual, respecting their unique abilities and preferences. Adaptability means that the technology can evolve with the user, adjusting to changes in their condition or their communication needs over time. Natural interaction focuses on making the communication process as intuitive and effortless as possible, minimizing the cognitive and physical load on the user. This means moving away from clunky, unintuitive interfaces towards systems that feel like natural extensions of the user's own voice and intent.

Achieving natural interaction also involves leveraging AI's capacity for understanding context and pragmatics. Communication is not just about words; it's about intent, tone, and the shared understanding between interlocutors. AI models are increasingly being developed to grasp these subtleties. For instance, an AI communication aid might understand that a user's request for "water" in a hot environment implies a need for a cold drink, or that a simple "yes" in response to a complex question might be a simplified affirmation due to communication fatigue. This deeper contextual understanding allows the AI to provide more relevant and nuanced communication outputs, making the user's interactions more effective and less prone to misinterpretation.

Social inclusion is a primary goal of these AI-driven communication aids. By empowering individuals to express themselves more freely and effectively, these tools break down social barriers and foster greater participation in community life, education, and employment. When individuals can articulate their thoughts, needs, and desires without significant impediment, their sense of self-worth and their ability to form meaningful relationships are profoundly enhanced. This is not merely about functional communication; it is about enabling full participation in the human experience. The ability to tell a joke, share an opinion, express love, or engage in a debate are fundamental aspects of human connection that these technologies aim to restore or enhance.

The ethical considerations in the development and deployment of these AI communication aids are also paramount. Ensuring data privacy and security is critical, as these systems often collect sensitive personal information about an individual's communication patterns and even their thoughts. Transparency in how AI models are trained and how data is used is essential to build trust with users and their families. Furthermore, there is a need to guard against over-reliance on technology, ensuring that it complements, rather than replaces, human interaction and support. The goal is to enhance a person's ability to connect with others, not to isolate them further through an overly technological interface. The development of these tools must be a collaborative process, involving not just AI engineers and designers, but also speech-language pathologists, occupational therapists, ethicists, and, most importantly, the individuals who will be using these technologies. Their lived experiences and insights are indispensable for creating tools that are truly empowering and beneficial. The continuous feedback loop from users is vital for iterative improvement, ensuring that these AI systems remain aligned with human needs and values. As AI technology continues to advance, we can anticipate even more sophisticated and personalized communication aids that will further democratize communication and foster a more inclusive society for all.

The integration of artificial intelligence into assistive technologies offers unprecedented opportunities to enhance the lives of individuals with disabilities. However, the very power and sophistication of these AI systems also introduce a critical ethical and practical imperative: the unwavering commitment to user control and agency. While AI can offer remarkable support by automating tasks, predicting needs, and facilitating communication, it must remain a tool that serves the user, not a system that dictates their actions or choices. The goal is to empower individuals, not to create an environment where they feel beholden to or superseded by the technology. This means designing systems that are not only intelligent and helpful but also transparent, adaptable, and, most importantly, subservient to the user's will.

A cornerstone of ensuring user control is the implementation of robust and intuitive override mechanisms. Users must always possess the ability to readily and easily disregard or correct an AI's suggestion, action, or prediction. This is particularly crucial in communication aids. If an AI predictive text system suggests a word or phrase that is inaccurate, inappropriate, or simply not what the user intended, there must be a straightforward way to dismiss the suggestion and input the correct information without undue effort or frustration. This could manifest as a simple swipe to dismiss, a dedicated "cancel" button, or even a gesture or vocal command that the AI is programmed to recognize as an explicit rejection. The speed and ease with which an override can be enacted are paramount. A complex or time-consuming override process can inadvertently create a sense of helplessness and dependency, defeating the purpose of assistive technology. Imagine a user trying to communicate an urgent need, and the AI repeatedly suggests incorrect words. If dismissing these suggestions is a multi-step process, the user's ability to convey their message efficiently is severely hampered, leading to anxiety and frustration. Therefore, these override functions should be as accessible and immediate as the AI's primary assistive functions.

Furthermore, the design of AI-powered assistive technologies must actively guard against the creation of user dependency. While the AI is intended to augment capabilities, it should not inadvertently erode a user's existing skills or their confidence in their own decision-making. For instance, in a navigation aid for visually impaired individuals, the AI might provide turn-by-turn directions. However, the system should also offer opportunities for the user to confirm their understanding of the route, perhaps by asking them to recall the next landmark or to indicate their confidence in the current path. This active engagement helps maintain the user's cognitive awareness of their surroundings and their ability to make independent judgments. Similarly, in adaptive learning systems, the AI might personalize educational content, but it should also include options for the user to deviate from the suggested path, explore topics of personal interest, or even challenge the AI's recommendations. This fosters a sense of ownership over the learning process and reinforces the user's role as the primary agent. The objective is to create a collaborative partnership between the user and the AI, where the AI acts as a sophisticated co-pilot, enhancing the journey without taking over the steering wheel.

Granular privacy settings are another non-negotiable aspect of maintaining user agency. Assistive technologies, especially those leveraging AI, often collect vast amounts of personal data to learn and adapt. This data can include communication patterns, daily routines, personal preferences, health information, and even emotional states. Users must have complete transparency and control over what data is collected, how it is used, and who it is shared with. This necessitates offering users a comprehensive suite of privacy controls that are easy to understand and manage. Instead of a one-size-fits-all approach to data collection, users should be able to opt in or out of specific data-gathering features. For example, a user might wish to allow their communication aid to learn their vocabulary for personal conversations but prefer not to share this data with third-party developers for product improvement. The settings should be presented in clear, non-technical language, avoiding jargon that could confuse or intimidate users. Moreover, users should be able to review the data that has been collected about them and have the right to request its deletion. This level of control not only respects user privacy but also builds trust, which is essential for the long-term adoption and effective use of assistive technologies.

Transparency in AI operations is equally vital for fostering user trust and control. Users should have a general understanding of how the AI makes its decisions or suggestions, even if they don't need to comprehend the intricate algorithms. This transparency can be achieved through various means. For example, when an AI communication aid suggests a word or phrase, it could briefly indicate *why* it made that suggestion. This might be a simple tag like "based on recent context" or "frequently used by you." This provides insight into the AI's reasoning process and allows the user to better evaluate the suggestion's relevance. Similarly, in adaptive interfaces, the AI might reconfigure the layout based on usage patterns. The system could offer a brief explanation, such as "I've moved your most used apps to this section for quicker access." This demystifies the changes and empowers the user to understand and even influence the adaptation process. When the AI's behavior is opaque, users are more likely to feel bewildered, frustrated, or even suspicious, which can undermine their sense of control and their willingness to rely on the technology.

The design process itself must be user-centric, with continuous involvement from the individuals who will use the assistive technology. This is not merely about testing a finished product; it's about co-designing the system from its inception. User feedback should actively shape the development roadmap, prioritizing features and functionalities that genuinely enhance autonomy and control. This involves engaging users in workshops, interviews, and usability studies, and creating feedback loops where their input leads to tangible improvements. For instance, if users consistently find a particular override mechanism cumbersome, designers must prioritize simplifying it. If users express a desire for more customizable prediction algorithms, the development team should explore ways to accommodate this. This collaborative approach ensures that the technology evolves in a way that aligns with the users' lived experiences and their fundamental need to remain in charge of their own lives.

Consider the nuanced challenges faced by individuals

with severe motor impairments who rely on eye-gaze technology for communication. While the AI can accurately track gaze and predict intended selections, accidental blinks, stray glances, or simply fatigue can lead to incorrect inputs. To ensure user control, these systems need sophisticated algorithms that can distinguish intentional selections from involuntary movements. Furthermore, users must have the ability to define a "dwell time" – the duration their gaze must remain on an item to register as a selection. This parameter should be highly customizable, allowing users to set it to a level that balances speed with accuracy, preventing both unintended activations and the frustration of having to hold their gaze for an uncomfortably long time. Moreover, there should be an easy way to "backspace" or correct a mistaken selection, perhaps through a specific eye movement or a secondary input device if available. The AI should also learn to interpret patterns of frustration – perhaps a user repeatedly selecting and deselecting an option – and offer assistance or a more intuitive pathway.

In the realm of AI-powered prosthetics and orthotics, user control extends to the physical interaction with the device. While AI can learn to anticipate movements and provide seamless assistance, the user must always feel like they are the one in command. This means ensuring that the prosthetic responds instantly and precisely to the user's intended actions, without lag or unexpected jolts. The AI should be trained on the user's specific movement patterns and intentions, allowing for personalized control. Moreover, users should have the ability to switch between different modes of operation or adjust the level of AI assistance on the fly. For example, someone using an AI-enhanced robotic arm might need to switch from a fine-motor task like picking up a small object to a gross-motor task like opening a door. The transition between these modes must be intuitive and user-driven, not dictated by the AI. The system could offer a menu of options that the user can select with minimal effort, or it could learn to infer the user's intention based on context and speed of movement. The ultimate goal is to create a symbiotic relationship where the technology amplifies the user's capabilities without ever compromising their sense of ownership and control over their own body and actions.

Another crucial area is the ethical consideration of AI in cognitive assistive technologies. For individuals with memory loss or executive function challenges, AI can provide prompts, reminders, and organizational support. However, it is imperative that these systems do not infantilize the user or undermine their self-efficacy. Instead of simply presenting a list of tasks, the AI could offer choices: "Would you like to tackle the email first, or start on the report?" or "Here are three options for what to do next, which one sounds most appealing?" This approach respects the user's autonomy and their right to make decisions, even if those decisions are supported by AI. The AI should act as a facilitator of decision-making, providing information and structure, rather than as an authority figure dictating tasks. Furthermore, users should have control over the frequency and style of reminders.

Someone might prefer gentle nudges, while another might need more direct prompts. The ability to adjust these settings ensures that the assistance is helpful without becoming intrusive or overwhelming.

When AI is used to analyze and interpret user behavior, such as through emotion detection or activity monitoring, the ethical implications for control and agency are particularly profound. While such technologies can be beneficial for understanding distress or identifying needs, they must be implemented with extreme caution and with explicit user consent and control. Users should have the ability to turn these monitoring functions on or off at any time and to understand what data is being collected and how it is being interpreted. For example, if an AI is designed to detect signs of depression, the user should be fully informed about this feature, and the AI's interpretation should always be presented as a suggestion for further discussion with a human professional, rather than a definitive diagnosis. The user must always retain the right to disagree with the AI's assessment and to control how such information is used or shared. Over-reliance on AI-driven interpretations of emotional states could lead to misinterpretations, stigmatization, and a reduction in the user's sense of self-understanding and agency.

Ultimately, the principle of user control and agency in AI-powered assistive technologies boils down to a fundamental respect for the individual's autonomy and dignity. The technology should be designed to augment human capabilities and to enhance the user's ability to live their life as they choose. This means prioritizing transparency, providing clear override mechanisms, offering granular privacy controls, and engaging users as active partners in the design and development process. When these principles are upheld, AI-powered assistive technologies can truly be transformative tools, empowering individuals to achieve greater independence, foster deeper connections, and lead more fulfilling lives. The technology should adapt to the user, not the other way around, ensuring that the human remains firmly in the driver's seat of their own journey.

The pervasive influence of artificial intelligence is reshaping numerous aspects of human life, and its integration into assistive technologies represents one of its most profound and ethically significant applications. While previous discussions have underscored the paramount importance of user control and agency in the design and deployment of these systems, a closer examination of specific applications reveals the tangible benefits and inherent challenges of AI-driven cognitive support. This case study delves into the practical implementation of AI technologies aimed at augmenting the cognitive abilities of individuals facing a spectrum of challenges, from attention deficits to memory impairments, and explores the delicate balance between providing essential support and respecting individual autonomy.

The realm of cognitive support is vast and diverse,

encompassing a range of conditions that can impact an individual's ability to manage daily life, maintain focus, and recall information. For individuals with Attention-Deficit/Hyperactivity Disorder (ADHD), the challenges often revolve around executive functions such as planning, organization, time management, and sustained attention. Similarly, individuals experiencing memory loss due to aging, neurological conditions like dementia, or traumatic brain injuries often struggle with recalling past events, remembering appointments, and navigating familiar environments. In both these scenarios, AI has emerged as a powerful ally, offering sophisticated yet often subtle interventions designed to bridge cognitive gaps and foster greater independence.

One of the most direct applications of AI in cognitive support lies in the development of intelligent reminder and scheduling systems. These are not merely digital calendars; they are dynamic entities that learn user patterns, anticipate needs, and offer contextualized prompts. For someone with ADHD, a typical day can be a minefield of distractions and forgotten tasks. An AI-powered assistant can help by breaking down larger goals into manageable steps, providing timed prompts for each stage of a task, and alerting the user if they deviate from their intended workflow. For example, if a user is working on a report, the AI might send a gentle reminder to focus, or suggest a short break after a predetermined period of intense activity, based on learned attention spans. These reminders are often designed to be non-intrusive, perhaps appearing as a discreet notification on a screen or a soft auditory cue, rather than an abrupt interruption that could exacerbate distraction. The system might also learn which types of prompts are most effective for a particular user – for instance, some might respond better to visual cues, while others require more direct verbal prompts.

For individuals with memory impairments, the need for reliable recall is critical for daily functioning and maintaining a sense of self. AI can serve as an external memory prosthesis, offering timely prompts for medications, appointments, or even remembering the names of people. Consider a person with early-stage dementia who might forget to take their daily medication. An AI system, integrated into a smart home environment or a wearable device, could not only remind them at the precise time but also confirm that the medication has been taken, perhaps by prompting them to interact with a sensor or answer a simple question. Beyond simple reminders, AI can also provide contextual recall. If a user is looking for a misplaced item, an AI system that has been passively observing their routines (with explicit consent, of course) might be able to offer suggestions: "You were in the kitchen around 10 AM, perhaps you left your keys there?" or "You often put your glasses on the bedside table when you go to sleep." This contextual information can be invaluable, reducing the frustration and anxiety often associated with memory loss.

The design of interfaces for these cognitive support

systems is a crucial element in their success. Overwhelming the user with too much information or too many options can be counterproductive, especially for individuals who are already experiencing cognitive overload. Therefore, human-centered AI design prioritizes simplicity, clarity, and contextual relevance. Interfaces are often minimalistic, employing clear typography, intuitive icons, and a limited color palette to reduce visual clutter. The information presented is typically filtered and prioritized based on the user's immediate needs. For instance, a daily schedule displayed by an AI assistant might initially show only the next upcoming event, with the option to reveal more details if requested. This "just-in-time" delivery of information prevents the user from being bombarded with a long list of future tasks, allowing them to focus on what is relevant in the present moment.

Personalization is another cornerstone of effective cognitive support. AI algorithms excel at learning individual patterns, preferences, and challenges. This learning allows the system to adapt its support strategies over time. For example, an AI might learn that a particular user responds best to reminders given 15 minutes before an event, rather than 5 minutes, or that they tend to forget tasks that are not scheduled before lunchtime. This adaptive capability ensures that the support remains relevant and effective, evolving alongside the user's needs and capabilities. The system can also learn to recognize signs of distress or confusion. If a user is repeatedly struggling with a particular task or seems disoriented, the AI could be programmed to offer more explicit guidance or to alert a caregiver or family member, if that has been pre-arranged and consented to.

A compelling area of AI application in cognitive support is the management of daily routines and task initiation for individuals with executive dysfunction. For someone struggling with initiation, the mere thought of starting a task can be paralyzing. AI can act as a gentle nudge or a structured guide. Imagine an AI system designed to help with morning routines. It might start by prompting the user to get out of bed at a specific time, then guide them through a sequence of actions: "Time to brush your teeth," followed by "Now, it's time to get dressed." Each step is presented clearly, and the AI can adapt the timing and sequence based on the user's progress and preferences. This structured approach can transform daunting tasks into a series of manageable, achievable steps, building confidence and fostering a sense of accomplishment.

The integration of AI into assistive technologies for cognitive support also necessitates careful consideration of ethical implications, particularly regarding privacy and autonomy. While these systems collect data to learn and adapt, the utmost care must be taken to protect user privacy. Consent must be explicit and ongoing, and users should have clear control over what data is collected, how it is used, and who has access to it. The data collected often includes highly personal information about daily habits, cognitive performance, and even emotional states. Transparency in how this data is handled is crucial for building trust. For example, if an AI system is used to track a user's focus during work, they should be fully aware of this monitoring and have the ability to disable it at any time.

Furthermore, the AI should always augment, not replace, the user's decision-making capabilities. Even when providing prompts and suggestions, the ultimate choice should remain with the individual. For instance, if an AI offers several options for an activity, it should present them neutrally, allowing the user to make the selection, rather than subtly steering them towards a predetermined choice. The goal is to empower individuals by providing them with the tools and support they need to manage their own lives, rather than creating a dependency on the technology that removes their agency. This means designing systems that are responsive to user input and allow for easy correction of errors or deviations from the AI's suggestions.

Consider the case of a person with early-onset Alzheimer's disease. Their ability to navigate their home, remember appointments, and maintain social connections can become increasingly challenging. An AI-powered system could be developed to assist them. This system might involve a combination of wearable sensors, smart home devices, and a user-friendly interface accessible via a tablet or smartphone. The AI could learn the individual's daily routine, such as when they typically wake up, eat meals, and engage in hobbies. Based on this learned routine, it could offer prompts. For example, a reminder to have breakfast, perhaps accompanied by a visual cue of a favorite meal. If the individual is planning to go out, the AI could provide a step-by-step guide for getting ready, including reminders to take their wallet and keys, and pre-programmed directions to their destination, with options to pause or repeat instructions. The AI could also be programmed to recognize signs of confusion or distress, such as repeated questioning or agitated behavior, and to offer calming interventions or alert a designated caregiver or family member, with prior consent.

The design of such a system would involve extensive user testing and co-creation with individuals experiencing cognitive decline, their families, and healthcare professionals. The interface would need to be exceptionally intuitive, perhaps utilizing large buttons, clear visual aids, and a voice-enabled interface that can understand natural language commands. The AI's responses would need to be patient and repetitive if necessary, without conveying frustration. Importantly, the system would need to incorporate robust privacy controls, allowing the individual and their trusted caregivers to manage data sharing and access permissions. The emphasis would always be on supporting independence and dignity, ensuring that the technology acts as a helpful companion rather than an intrusive overseer.

Another illustrative example is the application of AI in supporting individuals with Traumatic Brain Injury (TBI). Survivors of TBI often experience a range of cognitive deficits, including difficulties with attention, memory, problem-solving, and emotional regulation. AI-powered tools can provide targeted support for these challenges. For instance, an AI-driven "cognitive coach" application could help individuals manage their workload by breaking down tasks into smaller, more manageable steps, and providing reminders and prompts to stay on track. It might incorporate features that help users practice memory strategies, such as spaced repetition or mnemonic devices, and provide feedback on their progress. For individuals who struggle with time management, the AI could offer visual timers and alerts to help them allocate their time effectively.

A key design consideration for TBI survivors is the potential for sensory overload. Therefore, the AI interface should be highly customizable, allowing users to adjust the intensity of notifications, the complexity of information displayed, and the types of feedback they receive. Some individuals might benefit from a highly structured and predictable environment, while others might prefer more flexibility. The AI should be able to learn these preferences and adapt its behavior accordingly. Furthermore, the AI could be designed to detect early signs of cognitive fatigue or overwhelm, perhaps by monitoring interaction patterns or vocal cues, and suggest breaks or simpler activities. The aim is to create a supportive environment that minimizes cognitive load and promotes gradual recovery and increased self-efficacy.

The development of AI for cognitive support is not without its ethical quandaries. One significant concern is the potential for over-reliance on technology, which could inadvertently lead to a de-skilling of cognitive abilities or a reduction in an individual's intrinsic motivation. To mitigate this, the AI should be designed to foster active engagement and to encourage the user to take ownership of their cognitive processes. For example, instead of simply providing an answer, the AI might guide the user through a problem-solving process, asking them questions that prompt them to recall relevant information or to think critically. When providing reminders, the AI could offer choices: "Would you like to be reminded about this meeting in 30 minutes or an hour?" This empowers the user and reinforces their role as the agent in their own life.

Moreover, the integration of AI into the lives of individuals with cognitive challenges raises questions about personhood and autonomy. As AI systems become more sophisticated, the line between technological assistance and technological control can blur. It is imperative that the design and implementation of these systems are guided by principles that prioritize the dignity, autonomy, and self-determination of the individual. This means ensuring that the AI is always a tool that serves the user, rather than a system that dictates their behavior or limits their choices. The ultimate goal is to use AI to unlock potential, enhance independence, and improve the quality of life for individuals facing cognitive challenges, while always upholding their fundamental rights and their inherent worth. The ongoing dialogue between technologists, ethicists, healthcare professionals, and, most importantly, the users themselves, is critical to ensuring that AI in cognitive support evolves in a way that is both beneficial and ethically sound.

Chapter 13: The Future of Human-Centered AI Engineering

The trajectory of artificial intelligence is not a static one; it is a dynamic and accelerating evolution, constantly pushing the boundaries of what is possible and how humans engage with intelligent systems. As we look towards the future of human-centered AI engineering, several emergent trends promise to fundamentally reshape our interactions, demanding new approaches to design, development, and ethical consideration. These are not merely incremental improvements but rather paradigm shifts that will necessitate a proactive and deeply integrated human-centric philosophy to ensure that AI continues to serve humanity's best interests.

One of the most significant and rapidly developing frontiers is the convergence of AI with immersive technologies, particularly augmented reality (AR) and virtual reality (VR). These technologies, once largely confined to gaming and niche professional applications, are rapidly maturing and finding broader utility across diverse sectors, from education and training to healthcare and remote collaboration. The integration of AI into AR and VR environments offers unprecedented opportunities to create richer, more intuitive, and highly personalized human-AI experiences. Imagine an AR overlay that, powered by AI, not only identifies objects in a user's field of vision but also provides contextual information, historical data, or interactive instructions tailored to the user's skill level and immediate task. For instance, a mechanic could use AI-enhanced AR glasses to receive real-time guidance on complex repairs, with the AI analyzing sensor data from the machinery and projecting step-by-step instructions, highlighting critical components, and even predicting potential issues before they arise. This isn't just about displaying information; it's about creating an intelligent layer that augments human perception and capability in a seamless, context-aware manner.

In VR, AI can be employed to create highly realistic and responsive virtual environments and characters. This is particularly relevant for training simulations where AI-driven non-player characters (NPCs) can provide dynamic, adaptive scenarios that mirror real-world complexity. For example, medical students could practice complex surgical procedures on AI-powered virtual patients whose physiological responses are dynamically simulated, offering a learning experience far beyond static anatomical models. The AI can analyze the student's performance, identify areas for improvement, and adapt the difficulty or introduce unexpected complications to enhance learning outcomes. Beyond professional training, AI in VR can foster more empathetic interactions. Consider training programs designed to improve intergroup understanding or empathy towards individuals with disabilities. AI can generate diverse virtual personas exhibiting a wide range of behaviors and emotional responses, allowing users to engage in simulated interactions that foster deeper comprehension and perspective-taking. The challenge here lies in ensuring that these AI-driven simulations are not only realistic but also ethically sound, avoiding the propagation of stereotypes or biases within their generated content.

The rise of embodied AI, essentially intelligent agents that exist in the physical world through robotics, represents another profound shift in human-AI interaction. While robots have been a staple of industrial automation for decades, the advent of more sophisticated AI, particularly in areas like natural language understanding, computer vision, and reinforcement learning, is leading to the development of robots that can interact with humans and their environment in more nuanced and collaborative ways. These are not just tools; they are becoming partners. Think of advanced service robots in healthcare that can assist elderly individuals with daily tasks, provide companionship, and even monitor vital signs, all while adapting to the unique needs and personalities of their human charges. The AI powering these robots needs to possess a deep understanding of social cues, be capable of safe and intuitive physical interaction, and maintain a level of transparency about its actions and intentions.

Consider a domestic robot designed to assist with household chores. Such a robot would need to navigate complex home environments, identify and manipulate a variety of objects, and understand and respond to human commands, both verbal and gestural. AI would be crucial for its perception (understanding the environment), its planning (deciding how to complete a task), and its execution (physically performing the task). Crucially, for human-centered design, the AI must also incorporate social intelligence. It needs to understand when to interrupt, when to offer assistance proactively, and how to communicate its own status or limitations in a way that is easily understood by humans. The design of these robots moves beyond purely functional efficiency to encompass aspects of trust, comfort, and natural interaction. The physical form factor, the way it moves, its vocalizations, and its overall demeanor all contribute to how humans perceive and interact with it. Engineering these robots requires a multidisciplinary approach, blending robotics, AI, psychology, and design to create systems that are not only capable but also socially acceptable and beneficial.

The advancement of explainable AI (XAI) is poised to be a cornerstone in building trust and enabling effective human-AI collaboration, especially as AI systems become more complex and their decisions more impactful. For too long, many sophisticated AI models, particularly deep learning networks, have operated as "black boxes," making it difficult to understand precisely
why a particular decision was reached. This lack of transparency poses significant challenges for human oversight, debugging, and adoption, particularly in high-stakes domains like healthcare, finance, and autonomous systems. Emerging XAI techniques aim to shed light on these processes, providing insights into the reasoning behind AI outputs.

Explainable AI is not a monolithic field; it encompasses a range of approaches, from developing inherently interpretable models (like decision trees or linear regression) to creating post-hoc explanation methods for complex models. Post-hoc methods might include techniques that highlight the input features most influential in a model's decision, generate counterfactual explanations (e.g., "if this input had been slightly different, the outcome would have changed"), or provide visual representations of the model's internal workings. For example, in a medical diagnosis scenario, an XAI system could not only suggest a diagnosis but also highlight the specific visual features in an X-ray or MRI scan that led to that conclusion, along with confidence scores. This allows clinicians to critically evaluate the AI's recommendation, cross-reference it with their own expertise, and ultimately make a more informed decision.

The implications of XAI for engineering design are profound. Designers and engineers will need to develop interfaces that effectively communicate AI explanations to end-users, tailoring the level of detail and complexity to the user's domain knowledge and needs. This might involve creating interactive dashboards that allow users to probe the AI's reasoning, visualize its decision-making process, and even provide feedback that can refine future explanations. The challenge is to present explanations that are accurate, understandable, and actionable without overwhelming the user or leading to misinterpretations. A poorly explained AI decision can be more detrimental than no explanation at all. Therefore, the design of explanation interfaces requires as much rigor as the development of the AI models themselves, focusing on principles of clarity, conciseness, and user control.

The integration of these emerging trends – AI in AR/VR, embodied AI, and XAI – necessitates a fundamental rethinking of how we engineer AI systems. The focus must shift from simply building intelligent algorithms to creating intelligent systems that can seamlessly and ethically integrate into human lives. This requires a proactive, anticipatory approach to design, one that considers the full lifecycle of human-AI interaction and the potential societal impacts.

The engineering design practices of the future will undoubtedly be more interdisciplinary. The traditional silos between software engineering, hardware engineering, AI research, cognitive science, psychology, and ethics will continue to blur. Human-centered AI engineering will require teams composed of individuals with diverse expertise, all working collaboratively from the initial stages of ideation through to deployment and ongoing maintenance. Co-design methodologies, where end-users and stakeholders are actively involved in the development process, will become even more critical. This is especially true when developing embodied AI or AI integrated with immersive technologies, as the user experience is deeply intertwined with the physical and perceptual aspects of the interaction.

Furthermore, the iterative nature of AI development, particularly with machine learning, demands new approaches to system validation and verification. How do we ensure the safety, reliability, and fairness of AI systems that are constantly learning and evolving? For embodied AI, this involves rigorous testing in simulated and real-world environments, with a strong emphasis on safety protocols and fail-safe mechanisms. For AI in AR/VR, the focus might be on evaluating the cognitive load imposed on users, the potential for cybersickness, and the psychological impact of prolonged immersion. For XAI, validation will involve assessing the accuracy and utility of the explanations provided and ensuring they contribute to user trust and understanding rather than confusion.

The ethical considerations surrounding these emerging trends are paramount and will shape future engineering decisions. With embodied AI, questions of autonomy, accountability, and the potential for job displacement become more acute. How do we design robots that collaborate with humans without undermining human agency or creating new forms of dependency? Who is responsible when an autonomous robot causes harm? These are not merely philosophical debates; they are engineering challenges that require the development of robust ethical frameworks embedded directly into the AI's architecture and decision-making processes.

In AR/VR, the ethical landscape includes issues of privacy, data security, and the potential for addiction or psychological manipulation. AI systems that can observe and interpret user behavior within these immersive environments have access to unprecedented amounts of personal data. Engineering these systems requires stringent data protection measures, transparent data usage policies, and mechanisms for user control over data collection and sharing. The potential for creating highly persuasive or addictive virtual experiences also raises concerns about user well-being and requires careful consideration in the design of AI-driven content and interaction loops.

Explainable AI, while a step towards transparency, introduces its own set of ethical considerations. If explanations are not truly reflective of the AI's reasoning, they can create a false sense of security or understanding. Conversely, if explanations reveal sensitive information about the model's training data or architecture, they could pose security risks. The development of XAI must therefore be accompanied by robust guidelines for responsible disclosure and interpretation of AI reasoning. Ensuring that explanations are fair and unbiased, and do not inadvertently reinforce societal prejudices, is also a critical ethical imperative.

The future of human-centered AI engineering, therefore, is not just about technological advancement;

it is about the responsible and thoughtful integration of AI into the fabric of human life. It demands a commitment to continuous learning, adaptation, and a deep respect for human values and dignity. As we move forward, the ability to anticipate the human implications of new AI capabilities and to proactively design systems that foster trust, enhance human potential, and uphold ethical principles will be the defining characteristic of successful and beneficial AI engineering. The challenges are significant, but the opportunities to create a future where AI empowers and enriches human lives are even greater. The next wave of innovation will be defined by how well we can navigate this complex interplay between intelligent machines and the human beings they are designed to serve. The careful consideration of AI's role in AR/VR, the embodiment of AI in robotics, and the crucial development of explainable AI will collectively pave the way for a future where AI is not just powerful, but also profoundly human-aligned.

As artificial intelligence transcends its nascent stages and embeds itself ever more deeply within the fabric of our societies, the imperative to construct robust and adaptable ethical frameworks grows exponentially. The increasing autonomy of AI systems, coupled with their integration into critical infrastructure – from power grids and financial markets to healthcare diagnostics and autonomous transportation – amplifies the stakes of every decision an AI makes. These are no longer theoretical discussions confined to academic circles; they are pressing engineering challenges that demand immediate and sustained attention. The very definition of responsibility becomes a fluid concept when an AI system, composed of millions of lines of code and trained on vast, often opaque, datasets, makes a decision that leads to unintended consequences. Identifying the locus of accountability – whether it lies with the original developers, the deployers of the system, the data providers, or even the AI itself in a future of advanced artificial general intelligence – is a multifaceted problem that current legal and ethical structures are only beginning to grapple with.

The challenge of ensuring fairness in AI algorithms is equally intricate, particularly as these algorithms become more sophisticated and their decision-making processes less transparent. Bias, often inadvertently introduced through skewed training data or flawed algorithmic design, can perpetuate and even exacerbate existing societal inequalities. Consider AI systems used in hiring processes: if trained on historical data where certain demographic groups were underrepresented in specific roles, the AI might disproportionately filter out qualified candidates from those groups, thus reinforcing discriminatory patterns. The pursuit of fairness is not a simple binary; it involves navigating complex trade-offs between different notions of equity – equality of opportunity versus equality of outcome, for instance. Engineering fair AI requires not only technical solutions like bias detection and mitigation techniques but also a deep understanding of the socio-cultural context in which the AI will operate. This necessitates a proactive approach, embedding fairness considerations from the very conception of an AI system, rather than attempting to retroactively 'fix' it. This involves meticulous dataset curation, rigorous auditing of algorithmic outputs, and the development of metrics that can accurately assess fairness across diverse user groups. The complexity deepens when AI systems are designed to learn and adapt over time; what might be considered fair at one moment could become inequitable as the AI evolves, demanding continuous monitoring and recalibration of ethical parameters.

Furthermore, safeguarding human values in the face of increasingly powerful AI presents a fundamental philosophical and practical quandary for engineers. As AI systems become more capable of independent action and more adept at influencing human behavior – through personalized content recommendations, sophisticated persuasive technologies, or even simulated companionship – there is a growing risk of their goals diverging from or even undermining human well-being. The design of AI systems must therefore be guided by a clear set of human-centric values, such as autonomy, dignity, privacy, and flourishing. This requires a conscious effort to embed these values into the design process, not as an afterthought, but as core design principles. For instance, an AI designed to assist elderly individuals should prioritize their autonomy and dignity, offering support rather than dictating actions, and ensuring that it does not foster undue reliance or isolation. The development of AI in areas like mental health support also necessitates extreme caution, ensuring that therapeutic interventions are evidence-based, respectful of patient privacy, and do not create new forms of psychological dependency or harm. The challenge lies in translating abstract ethical principles into concrete engineering specifications and algorithmic constraints.

The dynamic and evolving nature of AI capabilities necessitates an equally dynamic and evolving approach to ethical frameworks. Static ethical guidelines, developed at a single point in time, are unlikely to remain relevant or effective as AI technologies advance at an unprecedented pace. This calls for a paradigm of continuous dialogue and collaboration, bridging the often-siloed disciplines of engineering, ethics, law, and social science. Engineers who build AI systems need to be equipped with a strong ethical literacy, understanding the potential societal implications of their work. Conversely, ethicists and policymakers need to gain a deeper appreciation for the technical realities and constraints of AI development to formulate practical and effective regulations.

This collaborative ecosystem is not merely an idealistic aspiration but a pragmatic necessity. Consider the development of AI for autonomous vehicles. Ensuring safety requires not only the engineering of reliable sensors and control systems but also robust ethical programming that can navigate complex moral dilemmas, such as deciding how to act in an unavoidable accident scenario. Such decisions, which involve trade-offs between different types of harm, cannot be left solely to engineers or even to a pre-programmed set of rules. They require broad societal consensus, informed by ethical deliberation and translated into technical requirements. This means that the process of developing ethical frameworks for AI must be inclusive, actively involving not only experts but also the general public. Public consultations, participatory design workshops, and accessible educational initiatives can help foster a shared understanding of AI's potential benefits and risks, and ensure that the AI systems we build reflect the values of the society they are intended to serve.

Moreover, the rapid pace of AI innovation means that regulatory frameworks must be agile and adaptable. Rigid, prescriptive regulations may quickly become obsolete, stifling innovation without effectively addressing emerging ethical concerns. Instead, a more flexible, principles-based approach, coupled with mechanisms for ongoing review and adaptation, may be more appropriate. This could involve establishing independent oversight bodies, promoting industry best practices, and fostering a culture of ethical responsibility within AI development organizations. The emphasis should be on creating an environment where ethical considerations are deeply integrated into the entire AI lifecycle, from research and development to deployment and ongoing maintenance.

The development of AI governance structures also needs to consider the global nature of AI. AI technologies transcend national borders, and the ethical challenges they pose are often international in scope.

This underscores the need for international cooperation in developing shared ethical principles and standards for AI. Harmonizing regulations and fostering cross-border collaboration can help prevent a 'race to the bottom,' where countries might relax ethical standards to gain a competitive advantage. It also ensures that the benefits of AI are shared equitably across the globe, and that its risks are managed collectively.

A critical aspect of building ethical frameworks for advanced AI is the concept of accountability. As AI systems become more sophisticated and their decision-making processes more intricate, assigning responsibility when things go wrong becomes increasingly challenging. In traditional engineering, accountability is often straightforward: if a bridge collapses due to faulty design, the engineers and the construction company are held liable. However, with AI, the lines of responsibility can become blurred. Who is responsible when an AI-driven trading algorithm causes a market crash, or when an AI diagnostic tool misses a critical illness? Is it the programmers who wrote the code, the company that deployed the system, the providers of the training data, or perhaps even the users who interacted with the AI in a way that led to the negative outcome? This ambiguity necessitates the development of new legal and ethical paradigms for AI accountability. One approach is to focus on establishing clear lines of responsibility throughout the AI development and deployment pipeline. This involves ensuring that every stage of the process is subject to rigorous oversight and that appropriate documentation is maintained. Another approach is to develop systems that can provide transparent audit trails of AI decision-making, allowing for post-hoc analysis to understand how a particular outcome was reached.

Furthermore, the concept of "meaningful human control" is emerging as a crucial element in ethical AI design, particularly for systems operating in high-stakes environments. This principle suggests that humans should retain the ability to understand, oversee, and, if necessary, intervene in the operation of AI systems. For example, in the context of autonomous weapons systems, the debate around meaningful human control is paramount. Critics argue that delegating lethal decision-making to machines poses unacceptable ethical risks, as AI systems may not possess the nuanced judgment, empathy, or understanding of proportionality required in complex combat situations. Engineers are therefore tasked with designing AI systems that augment, rather than replace, human decision-making, ensuring that a human remains "in the loop" or "on the loop" to exercise ultimate judgment. This has profound implications for interface design, requiring systems that can clearly communicate their reasoning, potential risks, and recommended actions in a timely and understandable manner, empowering human operators to make informed decisions.

The pursuit of fairness in AI also extends to ensuring that AI systems do not perpetuate or amplify existing societal biases. Machine learning algorithms learn from data, and if that data reflects historical discrimination, the AI will inevitably learn and replicate those biases. For instance, facial recognition systems have been shown to perform less accurately for individuals with darker skin tones or for women, due to biases in the datasets used for training. Addressing this requires a multi-pronged approach. Firstly, engineers must focus on creating diverse and representative datasets, actively seeking out and incorporating data from underrepresented groups. Secondly, they need to employ advanced algorithmic techniques for bias detection and mitigation, such as adversarial debiasing or re-weighting training samples. Thirdly, ongoing monitoring and evaluation of AI systems in real-world deployment are essential to identify and address any emergent biases that may not have been apparent during development. The challenge is that fairness itself is a complex and context-dependent concept. What constitutes "fair" treatment can vary significantly across different cultures and applications, and there can be inherent trade-offs between different metrics of fairness. For example, optimizing for equal opportunity might lead to different outcomes than optimizing for equal representation. Therefore, engineers must engage in careful ethical deliberation to determine which definition of fairness is most appropriate for a given application and to be transparent about the trade-offs involved.

Safeguarding human values in the age of advanced AI requires a proactive embedding of ethical principles into the very architecture of AI systems. This is often referred to as "ethics by design" or "value-sensitive design." It means that ethical considerations are not an add-on or a compliance checklist, but rather an integral part of the design process from its inception. For example, when designing AI for personalized education, engineers might incorporate features that promote student autonomy and intrinsic motivation, rather than solely optimizing for standardized test scores. This could involve providing students with choices in their learning paths, offering constructive feedback that encourages self-reflection, and ensuring that the AI does not create an environment of excessive pressure or anxiety. Similarly, AI systems intended for social interaction should be designed to promote genuine connection and respect, rather than manipulative engagement or the creation of echo chambers. This requires careful consideration of the AI's conversational style, its response mechanisms, and its ability to understand and respond to nuanced human emotions. It also entails designing in safeguards against the misuse of AI for manipulation or disinformation campaigns.

The ongoing dialogue needed to shape responsible AI development must extend beyond the technical and academic communities to encompass a broader societal conversation. Policymakers have a critical role to play in establishing clear regulatory frameworks that can guide AI development and deployment, ensuring accountability and protecting fundamental human rights. However, these regulations must be informed by a deep understanding of the technology and its potential impacts, and they must be flexible enough to adapt to the rapid pace of innovation. Industry leaders and companies developing AI have a responsibility to foster a culture of ethical awareness and to invest in responsible AI practices, including robust internal review processes, ethical training for their employees, and transparency with the public about their AI systems. Researchers, in turn, must continue to explore the ethical implications of new AI capabilities and to develop innovative solutions for addressing them.

Ultimately, building ethical frameworks for advanced AI is not a one-time task but an ongoing process of adaptation, learning, and collaboration. It requires a commitment to continuous critical reflection, a willingness to engage with diverse perspectives, and a shared understanding that the ultimate goal of AI development must be to serve humanity and to enhance human well-being, rather than to pursue technological advancement for its own sake. The future of human-centered AI engineering hinges on our collective ability to navigate these complex ethical landscapes with foresight, integrity, and a steadfast dedication to human values. The intricate interplay between increasingly capable AI systems and the human societies they inhabit demands a proactive, collaborative, and ethically grounded approach to engineering that prioritizes not just what AI *can* do, but what it *should* do. This necessitates a sustained commitment to interdisciplinary dialogue, involving engineers, ethicists, social scientists, policymakers, and the public, to ensure that the powerful tools we are creating are aligned with our

highest aspirations for a just, equitable, and flourishing future. The continuous refinement of our ethical compass, coupled with pragmatic engineering solutions, will be the hallmark of successful and beneficial AI integration.
The imperative to cultivate a new breed of AI engineers and designers, deeply attuned to human needs and ethical considerations, underscores the critical role of education and training. As AI systems become more pervasive and influential, the responsibility for their benevolent development and deployment rests squarely on the shoulders of those who architect them. This necessitates a fundamental reimagining of our educational pathways, moving beyond purely technical instruction to embrace a holistic approach that integrates ethical reasoning, user empathy, and collaborative practice.

The current landscape of AI education, while rapidly evolving, often remains heavily focused on algorithmic proficiency, data science techniques, and computational power. While these are undeniably essential, they represent only one facet of the required skillset for creating truly human-centered AI. A significant gap exists in preparing practitioners to grapple with the nuanced ethical dilemmas, the potential societal impacts, and the diverse needs of end-users. Therefore, a crucial evolution in AI engineering education involves the deliberate and robust integration of AI ethics as a core component of curricula at all levels, from undergraduate programs to postgraduate specializations and professional development courses. This integration should not be a mere add-on, a single elective course that can be easily overlooked, but rather a pervasive theme woven through the entire educational fabric. Students should be exposed to foundational ethical theories, critically analyze case studies of AI failures and successes from an ethical perspective, and engage in rigorous discussions about the trade-offs inherent in AI design decisions. This means delving into concepts like fairness, accountability, transparency, privacy, and the prevention of harm, not as abstract philosophical exercises, but as practical engineering challenges that require tangible solutions.

Complementing ethical training, the principles and methodologies of user-centered design (UCD) must be elevated to a paramount position in AI engineering education. Too often, AI systems are developed with a "technology-first" mentality, where the capabilities of the algorithms dictate the product, and user needs are considered only as an afterthought, if at all. Human-centered AI engineering, however, demands that the user be at the very heart of the design process. This requires equipping future engineers with a deep understanding of UCD methodologies, including user research techniques such as interviews, surveys, and ethnographic studies, to genuinely understand user needs, contexts, and pain points. Furthermore, they must be proficient in iterative design processes, prototyping, and user testing to ensure that the AI systems they build are not only functional but also usable, desirable, and truly beneficial to the people they are intended to serve. This includes cultivating empathy

– the ability to step into the shoes of diverse user populations, recognizing that "users" are not a monolithic entity but a spectrum of individuals with varied backgrounds, abilities, and cultural contexts. Training should emphasize designing for inclusivity, accessibility, and dignity, ensuring that AI systems do not inadvertently marginalize or disadvantage any group.

The inherent complexity and multidisciplinary nature of AI development also underscore the vital need for interdisciplinary collaboration within educational settings. AI is not solely a computer science problem; it is deeply intertwined with sociology, psychology, philosophy, law, economics, and domain-specific expertise in areas like healthcare, finance, and education. Therefore, AI engineering programs should actively foster collaboration between students from different disciplines. This could manifest through joint project work, shared coursework, or even dedicated interdisciplinary AI labs. By working alongside peers from diverse backgrounds, future engineers will gain invaluable insights into the broader societal implications of their work and learn to communicate effectively across different professional languages and perspectives. This collaborative environment mirrors the realities of professional AI development, where successful teams are rarely composed of individuals with identical skill sets. It prepares students to engage constructively with ethicists, social scientists, legal experts, and domain specialists, fostering a more robust and well-rounded approach to AI creation.

Moreover, the relentless pace of technological advancement in AI mandates a paradigm of lifelong learning and continuous professional development. The skills and knowledge that are cutting-edge today can quickly become obsolete. Therefore, educational institutions and industry leaders must cultivate a culture where learning is not confined to formal schooling but is an ongoing, active pursuit throughout a practitioner's career. This involves encouraging and supporting employees in attending workshops, conferences, and advanced training programs focused on emerging AI technologies, ethical best practices, and new design methodologies. It also means creating internal knowledge-sharing platforms, mentorship programs, and opportunities for cross-pollination of ideas within organizations. The ability to adapt, unlearn, and relearn will be a defining characteristic of successful AI engineers in the future. Educational institutions can play a role here by offering modular courses, micro-credentials, and flexible online learning options that allow professionals to update their skills as needed without requiring a full return to traditional academic settings.

The development of AI ethics modules within engineering curricula should go beyond mere theoretical discussions. Practical exercises are crucial. This could involve students participating in ethical hacking scenarios tailored for AI, where they attempt to identify and exploit vulnerabilities that could lead to biased or harmful outcomes. They might be tasked with auditing existing AI systems for fairness and transparency, and proposing mitigation strategies. Furthermore, case studies should not shy away from real-world controversies, such as algorithmic discrimination in loan applications or biased facial recognition systems, prompting students to dissect the root causes and brainstorm engineering solutions. The goal is to move from simply understanding ethical principles to actively applying them in concrete design and development contexts. This hands-on experience is invaluable for building the practical judgment and ethical resilience required in the field.

The integration of user-centered design principles must also be translated into practical, project-based learning. Students should be encouraged to undertake projects that involve direct engagement with potential users from the outset. This might mean developing a prototype AI assistant for a specific community group, conducting user interviews to refine its functionalities, and iterating on the design based on their feedback. Such projects would naturally expose students to the challenges of data collection with diverse populations, the importance of clear and intuitive user interfaces for AI, and the ethical considerations surrounding data privacy and consent when interacting with real users. The emphasis would shift from building a technically complex algorithm to building a technology that genuinely solves a problem for a specific group of people in a way that respects their needs and autonomy. This practical application reinforces the idea that a technically perfect AI that is unusable or undesirable is a failed AI.

Furthermore, fostering interdisciplinary collaboration requires more than just co-locating students from different departments. It necessitates structured opportunities for them to work together on meaningful projects that require the integration of diverse expertise. For example, a team might consist of computer science students focusing on the AI algorithms, sociology students analyzing the potential societal impact and user demographics, psychology students designing user interfaces for optimal engagement, and even law students considering the regulatory and privacy implications. These collaborative projects should be assessed not only on the technical merit of the AI but also on the team's ability to integrate different perspectives, manage interdisciplinary communication, and arrive at a holistic solution that addresses technical, ethical, and social dimensions. This cultivates the soft skills – communication, negotiation, conflict resolution – that are essential for effective teamwork in complex, multi-stakeholder environments.

The concept of "value-sensitive design" should also be a cornerstone of AI engineering education. This approach explicitly considers human values throughout the design process, aiming to embed them into the technology itself. Educational programs should teach students how to identify relevant values for a given AI application (e.g., autonomy, privacy, fairness, safety, well-being) and how to translate these abstract values into concrete design features and technical constraints. For instance, when designing an AI that provides personalized news recommendations, value-sensitive design would prompt engineers to consider how to promote exposure to diverse viewpoints (counteracting echo chambers) and how to ensure transparency in the recommendation algorithm, thus upholding values of informed decision-making and intellectual freedom. This requires a shift in mindset from simply optimizing for engagement metrics to actively designing for a desired set of positive human outcomes.

Continuous professional development in AI ethics and human-centered design is not merely about keeping up with technical trends but also about evolving ethical sensibilities. As AI capabilities expand and societal contexts change, new ethical challenges will inevitably emerge. Therefore, training programs must be dynamic and responsive. This could include regular workshops on emerging AI ethical issues, such as the implications of generative AI for misinformation, the ethical considerations of AI in warfare, or the impact of AI on labor markets. Industry associations and professional bodies have a crucial role to play in developing and disseminating such training, establishing clear ethical guidelines and best practices, and fostering a community of practice where practitioners can share experiences and learn from one another. The emphasis must be on building a culture of continuous ethical reflection and improvement within AI development organizations.

Ultimately, the success of human-centered AI engineering hinges on our ability to equip current and future practitioners with the right blend of technical prowess, ethical acumen, and a deep-seated commitment to human well-being. This requires a concerted effort to transform educational paradigms, fostering interdisciplinary collaboration, embedding user-centered design and ethical considerations at every stage, and committing to a future of lifelong learning and adaptation. The engineers and designers we train today will be the architects of tomorrow's AI, and it is incumbent upon us to ensure they are equipped not only to build powerful systems but also to build them wisely, ethically, and with humanity at their core. This transformation in education is not an optional add-on; it is the foundational requirement for realizing the immense potential of AI for good, while mitigating its inherent risks and ensuring that these powerful technologies serve the best interests of all humanity. The development of these specialized curricula and training programs is a critical investment in a future where AI and humanity can coexist and thrive harmoniously. It is about cultivating a generation of problem-solvers who understand that the most elegant technical solution is often one that is also ethically sound and deeply considerate of the human experience.

The trajectory of human-centered AI engineering is inextricably linked to its capacity for fostering genuine and effective interdisciplinary collaboration. As we move beyond the realm of purely technical problem-solving, the inherent complexity of AI systems, their profound societal impact, and the paramount importance of human values necessitate a radical departure from siloed approaches. The creation of AI that is not only intelligent but also ethical, equitable, and truly beneficial to humanity demands a symphony of expertise, where engineers, designers, social scientists, ethicists, legal scholars, and domain specialists converge with a shared vision and a common language. This section delves into the critical imperative for such collaboration, exploring its foundational principles, practical manifestations, and the transformative potential it holds for the future of AI development.

The traditional model of engineering, often characterized by distinct departmental boundaries and specialized skill sets, is insufficient for the challenges posed by advanced AI. An algorithm that optimizes for efficiency might inadvertently perpetuate societal biases, an interface designed for ease of use could compromise user privacy, and a system deployed in a sensitive domain like healthcare requires an intimate understanding of both clinical practice and patient well-being. These are not merely technical hurdles; they are complex socio-technical problems that demand a confluence of perspectives. For instance, an AI system designed for personalized education, while technically brilliant in its adaptive learning algorithms, could fail to address the diverse learning styles, cultural backgrounds, and socio-economic disparities of students if the development team lacks input from educational psychologists, sociologists, and educators themselves. Similarly, an AI intended for judicial assistance must be rigorously vetted not only for its accuracy in processing legal precedents but also for its fairness and impartiality, requiring close collaboration with legal scholars, ethicists, and domain experts in jurisprudence.

At its core, fostering interdisciplinary collaboration means dismantling the walls that separate different fields of knowledge and practice. It requires cultivating an environment where individuals from diverse backgrounds feel empowered to contribute their unique insights and where their perspectives are valued and integrated into the design process. This is not simply about having representatives from different disciplines present in meetings; it is about creating a truly synergistic environment where shared ownership of the problem and the solution is paramount. Engineers need to understand the ethical frameworks that guide societal norms, designers must grasp the computational constraints and possibilities, social scientists need to articulate the nuanced ways in which technology interacts with human behavior and social structures, and ethicists and legal experts must translate abstract principles into actionable engineering considerations.

One of the fundamental challenges in achieving effective

interdisciplinary collaboration lies in overcoming differences in terminology, methodology, and even fundamental assumptions. A computer scientist's definition of "bias" might differ significantly from that of a sociologist, and an ethicist's approach to risk assessment may not align with a traditional engineering risk matrix. Therefore, a crucial element in building collaborative AI teams is the deliberate development of a shared lexicon and a common understanding of key concepts. This can be achieved through dedicated workshops, shared learning modules, and the creation of interdisciplinary "translation" roles within teams. The goal is to move towards a common ground where ideas can be communicated clearly and unambiguously, enabling a more fluid exchange of knowledge and a deeper mutual respect for each discipline's contributions.

The practical implementation of human-centered AI development offers fertile ground for interdisciplinary endeavors. Consider the development of an AI-powered diagnostic tool for a specific medical condition. The engineering team might focus on the machine learning models, data processing pipelines, and algorithmic accuracy. However, to ensure it is human-centered, this technical core must be augmented by several other disciplines. Medical professionals, such as doctors and nurses, are essential for providing context on clinical workflows, understanding patient needs, and identifying potential risks in real-world medical scenarios. Ethicists would guide the team in navigating issues of patient privacy, data security, algorithmic fairness in diagnosis across different demographic groups, and the ethical implications of delegating diagnostic decisions to an AI. Sociologists and anthropologists could offer insights into how such a tool might be adopted or rejected by different patient populations, understanding cultural variations in health beliefs and practices. UX/UI designers would ensure the tool is intuitive and easy to use for healthcare providers, considering factors like cognitive load and information display under pressure. Legal experts would advise on regulatory compliance, liability, and the legal framework governing medical devices. Without this integrated approach, a technically sound diagnostic AI could be unusable, inequitable, or even harmful in practice.

Another compelling example is the creation of AI systems for urban planning and smart city initiatives. An AI designed to optimize traffic flow, manage energy consumption, or predict public safety needs requires input far beyond that of traffic engineers and data scientists. Urban planners bring an understanding of city infrastructure, zoning laws, and the long-term spatial and social impacts of development. Environmental scientists are crucial for assessing the ecological footprint of AI-driven solutions. Sociologists and community organizers can provide vital perspectives on how these technologies might affect different neighborhoods, social equity, and citizen engagement. Policymakers and legal experts are needed to navigate the complex regulatory landscape and ensure that AI deployments align with public interest and legal frameworks. Furthermore, citizen participation, facilitated by communication experts and participatory design specialists, becomes an integral part of the process, ensuring that the "smart city" truly serves its inhabitants. The iterative development of such systems, informed by continuous feedback from diverse stakeholders, is a testament to the power of interdisciplinary collaboration in addressing complex societal challenges.

To facilitate this collaborative spirit, educational institutions and professional organizations play a pivotal role. The traditional academic structure, often rigidly compartmentalized, needs to evolve to encourage cross-pollination of ideas and skills. This can involve creating joint degree programs, interdisciplinary research centers, and collaborative project incubators where students and researchers from different fields can work together from the outset. For instance, a "Center for Responsible AI" might house computer scientists, ethicists, legal scholars, and social scientists under one roof, fostering a culture of shared inquiry and problem-solving. Projects within these centers could be structured to explicitly require teams to integrate diverse expertise, with project evaluations that assess not only the technical outcomes but also the effectiveness of the interdisciplinary collaboration itself.

In the professional sphere, organizations must actively cultivate an inclusive and collaborative culture. This means moving beyond functional silos and encouraging the formation of cross-functional teams for AI development projects. Leaders need to champion interdisciplinary work, providing the necessary resources, time, and structural support for collaboration to flourish. This could involve establishing common project management frameworks that accommodate diverse work styles, creating shared workspaces that encourage informal interaction, and implementing performance review systems that recognize and reward contributions to interdisciplinary team success. Furthermore, investing in training that enhances communication, negotiation, and conflict resolution skills across disciplines is essential for navigating the inevitable complexities of diverse team dynamics.

The development of "AI ethics boards" or "responsible AI governance committees" within organizations is another avenue for structured interdisciplinary collaboration. These bodies, comprising ethicists, legal counsel, domain experts, user advocates, and engineers, can provide oversight and guidance throughout the AI lifecycle, from initial concept to deployment and ongoing monitoring. Their mandate would be to scrutinize AI systems for potential risks, ensure alignment with ethical principles and societal values, and provide recommendations for mitigation and improvement. This institutionalizes the integration of diverse perspectives, ensuring that ethical and human-centered considerations are not an afterthought but a core part of the AI development process.

The challenge of building truly human-centered AI is not merely one of technical sophistication; it is fundamentally a challenge of human understanding and integration. It requires us to recognize that the most effective solutions will emerge from the synthesis of diverse intelligences, experiences, and perspectives. By actively promoting and structuring interdisciplinary collaboration, we can move closer to realizing the immense potential of AI to address humanity's most pressing challenges, ensuring that these powerful technologies are developed and deployed not just effectively, but also equitably, ethically, and with a profound respect for human dignity and well-being. This commitment to collaboration is not an optional extra; it is the very foundation upon which the future of human-centered AI engineering must be built. It represents a paradigm shift from isolated technical achievement to collective, values-driven innovation, where the ultimate measure of success is the positive and equitable impact on human lives and societies.

Moreover, the continuous evolution of AI necessitates a dynamic and responsive approach to interdisciplinary collaboration. As new AI capabilities emerge, such as advanced generative models or increasingly sophisticated reinforcement learning agents, they bring with them novel ethical and societal questions. For instance, the rise of powerful AI language models raises complex issues around authorship, intellectual property, misinformation, and the future of creative industries. Addressing these requires collaboration between AI researchers, linguists, philosophers, legal scholars specializing in intellectual property, media studies experts, and even artists and writers themselves. Similarly, the application of AI in areas like autonomous vehicles or advanced robotics necessitates close partnership with ethicists, legal experts specializing in tort law, cognitive psychologists studying human-robot interaction, and human factors engineers to ensure safety and societal acceptance. The collaborative structures must be agile enough to incorporate emerging expertise and adapt to new challenges as they arise.

The concept of "value-sensitive design," which explicitly integrates human values into the design process, is inherently interdisciplinary. It requires a deep understanding of which values are relevant, how they can be translated into design requirements, and how their presence or absence can be evaluated. This process naturally draws upon the insights of philosophers, sociologists, psychologists, and ethicists, who can help identify and articulate core human values such as autonomy, privacy, fairness, safety, and well-being. Engineers and designers then translate these abstract values into concrete technical specifications and design features. For example, when designing an AI system that manages energy distribution in a smart grid, engineers might focus on efficiency, but ethicists and social scientists would ensure that considerations of equitable access to energy, particularly for vulnerable populations, are embedded into the system's logic. This requires ongoing dialogue and co-creation between technical teams and value-holders.

Furthermore, for AI systems to be truly human-centered, they must be designed with and for diverse user groups. This often means engaging with communities that have historically been marginalized or underserved by technology. Building trust and ensuring meaningful participation from these communities is a complex undertaking that requires the expertise of social scientists, community organizers, and individuals with deep cultural understanding. Collaborative efforts might involve co-design workshops, participatory action research, and the establishment of community advisory boards. The insights gained from these direct engagements are invaluable for identifying potential biases, unintended consequences, and ensuring that AI solutions are relevant, accessible, and empowering for all. This moves beyond simply "understanding" user needs to actively involving users as co-creators in the development process.

The legal and regulatory landscape surrounding AI is also rapidly evolving, and effective AI development requires close collaboration with legal experts. Issues of data privacy, algorithmic accountability, liability for AI-driven harms, and the ethical use of AI in various sectors (e.g., finance, healthcare, employment) all demand legal scrutiny. Engineers and developers need to work hand-in-hand with lawyers to ensure compliance with existing regulations and to anticipate future legal frameworks. This collaboration helps in designing systems that are not only technically sound but also legally defensible and ethically compliant, mitigating risks of litigation and ensuring responsible innovation.

In conclusion, the future of human-centered AI engineering is inextricably tied to the depth and breadth of its interdisciplinary collaborations. Moving beyond the confines of traditional disciplinary silos is not merely a matter of convenience but an absolute necessity for developing AI that is intelligent, ethical, equitable, and truly serves humanity. By actively fostering environments that encourage shared understanding, mutual respect, and co-creation among engineers, designers, social scientists, ethicists, legal experts, and domain specialists, we can unlock AI's potential to solve complex global challenges and build a future where technology and humanity thrive in harmony. The investment in cultivating these collaborative capabilities is an investment in a more responsible, beneficial, and humane technological future.

The journey we have undertaken throughout this book has illuminated the profound potential of Artificial Intelligence, but more critically, it has underscored the essential responsibility that accompanies this power. We have explored the intricacies of human-centered AI engineering, not as a mere technical pursuit, but as a deeply humanistic endeavor. The algorithms we design, the systems we build, and the interfaces we craft have far-reaching implications, touching lives, shaping societies, and influencing the very fabric of our future. This is not a future that will unfold passively; it is a future that we must actively, consciously, and collaboratively shape. Therefore, this concluding section serves as a potent call to action, a resounding invitation for each of you – whether you are an engineer meticulously crafting code, a designer envisioning user experiences, a student embarking on your professional journey, or a seasoned professional navigating complex projects – to become architects of an ethical and human-centered AI future.

The core message that must resonate from our discussions is the unwavering imperative to place human well-being, fairness, and societal benefit at the vanguard of all our AI endeavors. This principle is not a peripheral consideration, a checkbox to be ticked, or a mere footnote in a project proposal. It is the foundational bedrock upon which all responsible AI development must stand. When we speak of "human-centered," we mean understanding that AI is not an end in itself, but a tool, an extension of human capability, designed to augment, assist, and ultimately, to uplift the human condition. This demands a profound empathy, a deep-seated understanding of human needs, aspirations, and vulnerabilities, and a commitment to ensuring that our technological creations serve these fundamental aspects of our existence.

Consider the myriad ways AI is already interwoven into

our lives. From the personalized recommendations that guide our consumption to the diagnostic tools assisting medical professionals, AI is no longer a futuristic concept but a present reality. Each application, however seemingly benign, carries with it the potential for both immense good and unintended harm. An AI designed to optimize resource allocation in a city, for instance, could inadvertently exacerbate existing inequalities if its underlying data reflects historical biases or if its optimization goals do not account for the needs of marginalized communities. Similarly, an AI employed in hiring processes, while aiming for efficiency, could perpetuate discrimination if it learns from past, biased hiring decisions. These are not hypothetical scenarios; they are tangible risks that demand our vigilant attention and proactive mitigation.

Our collective responsibility, therefore, extends beyond the creation of intelligent systems. It extends to the creation of
wise systems, systems that are not only capable but also considerate, not only efficient but also equitable, not only powerful but also principled. This necessitates a deliberate shift in our mindset, moving from a purely utilitarian approach to one that is deeply rooted in ethical reasoning and societal impact assessment. This means asking critical questions at every stage of the AI lifecycle: Who benefits from this technology? Who might be harmed? Are there unintended consequences we haven't foreseen? How can we ensure fairness and avoid perpetuating or amplifying existing societal

biases? What are the implications for individual autonomy and collective well-being?

This call to action is addressed to the engineers who write the algorithms, for they hold the power to embed fairness and accountability into the very logic of AI. It is for the designers who craft the interfaces, for they can create systems that are transparent, explainable, and that empower users rather than obfuscate or control. It is for the project managers and leaders who set the direction and allocate resources, for they have the crucial role of fostering a culture that prioritizes ethical considerations and interdisciplinary collaboration. And it is for the students and aspiring professionals, for you are the next generation of AI creators, and your commitment to these principles will define the trajectory of this transformative technology.

Embracing this responsibility requires a proactive and continuous learning posture. The field of AI is evolving at an unprecedented pace, and with each new advancement come new ethical challenges and societal considerations. Staying abreast of these developments, engaging with ethical frameworks, and fostering a spirit of intellectual curiosity are not optional extras; they are essential components of responsible AI engineering. This might involve seeking out specialized training in AI ethics, participating in industry forums that discuss responsible AI practices, or engaging in critical discourse with colleagues and peers about the ethical implications of our work. It means recognizing that ethical AI is not a static destination but an ongoing process of reflection, adaptation, and improvement.

Furthermore, this call to action strongly advocates for the integration of diverse perspectives. As we have explored, no single discipline possesses all the answers to the complex socio-technical challenges posed by AI. Engineers need to work hand-in-hand with ethicists, social scientists, legal scholars, domain experts, and, crucially, with the end-users themselves. This interdisciplinary approach is not merely about mitigating risks; it is about unlocking innovation. By bringing together a spectrum of viewpoints, we can identify novel solutions, anticipate a wider range of potential impacts, and develop AI systems that are more robust, inclusive, and ultimately, more beneficial to society. This means actively seeking out and valuing contributions from individuals with different backgrounds, experiences, and expertise, fostering an environment where open dialogue and constructive disagreement are encouraged.

The concept of transparency and explainability in AI systems is another critical element of this call to action. As AI becomes more sophisticated, its decision-making processes can become increasingly opaque, leading to a "black box" phenomenon. This lack of transparency erodes trust and makes it difficult to identify and rectify errors or biases. We must therefore commit to developing AI systems that are as interpretable as possible, providing users and stakeholders with a clear understanding of how decisions are made, what data is used, and what the potential limitations are. This not only builds trust but also empowers individuals to challenge AI-driven decisions and to hold developers accountable.

Accountability is a cornerstone of ethical AI. When AI systems make mistakes or cause harm, there must be clear lines of responsibility. This requires establishing robust governance frameworks, auditing mechanisms, and recourse procedures. It means moving away from a culture of "move fast and break things" towards one that prioritizes safety, reliability, and accountability. This might involve implementing rigorous testing protocols, establishing independent oversight bodies, and ensuring that mechanisms are in place for redress when AI systems fail to meet ethical or performance standards. The challenge lies in defining accountability in distributed systems where multiple actors and algorithms may contribute to an outcome.

Our commitment to an ethical AI future also demands a conscious effort to combat bias. AI systems learn from data, and if that data reflects historical societal biases – whether related to race, gender, socioeconomic status, or any other demographic factor – the AI will inevitably learn and perpetuate those biases, often at scale. Actively identifying, measuring, and mitigating bias in data and algorithms is a non-negotiable aspect of responsible AI development. This requires careful data curation, the development of bias detection tools, and the implementation of fairness-aware machine learning techniques. It also necessitates a critical examination of the very metrics we use to define "fairness," recognizing that different notions of fairness may be appropriate in different contexts.

Beyond addressing existing biases, we must also be mindful of the potential for AI to create new forms of inequality or to widen existing societal divides. The "digital divide" could be exacerbated if access to advanced AI technologies and the benefits they confer is unevenly distributed. We must therefore advocate for equitable access to AI education, resources, and applications, ensuring that the transformative potential of AI is shared broadly across all segments of society. This might involve initiatives to promote AI literacy in underserved communities, investments in public infrastructure that supports AI deployment, and policies that encourage the development of AI solutions for public good.

Moreover, the very definition of what it means to be human and how we interact with technology is being reshaped by AI. As AI systems become more sophisticated in their ability to understand and generate human language, to create art, or to provide companionship, we must grapple with the ethical implications for human relationships, creativity, and identity. A human-centered approach means ensuring that AI enhances, rather than diminishes, our humanity, fostering genuine connection and creativity rather than artificial replication or alienation. This requires ongoing philosophical and ethical reflection, alongside technical development.

This call to action is not about stifling innovation; it is about guiding it. It is about ensuring that the incredible ingenuity and problem-solving power of AI are harnessed for the greatest good, not for fleeting profit or unchecked technological advancement. It is about building a future where AI is a force for empowerment, equity, and human flourishing. This requires a sustained commitment from individuals, organizations, and governments to create an ecosystem that supports and incentivizes responsible AI development and deployment.

Ultimately, shaping an ethical AI future is a collective endeavor. It requires us to move beyond theoretical discussions and engage in practical implementation. It means integrating ethical considerations into every phase of the AI lifecycle, from initial conceptualization and data collection to model development, deployment, and ongoing monitoring. It means fostering a culture of continuous learning, open dialogue, and mutual accountability. It means recognizing that the future of AI is not predetermined; it is a choice we make every day through our actions, our decisions, and our commitments. Let us choose to build an AI future that reflects our highest values, a future that is not only intelligent but also wise, not only innovative but also inclusive, and not only powerful but profoundly humane. The time to act is now. Let us embrace this challenge and opportunity with open minds, a commitment to ethical principles, and a shared vision for a better, AI-enhanced world.

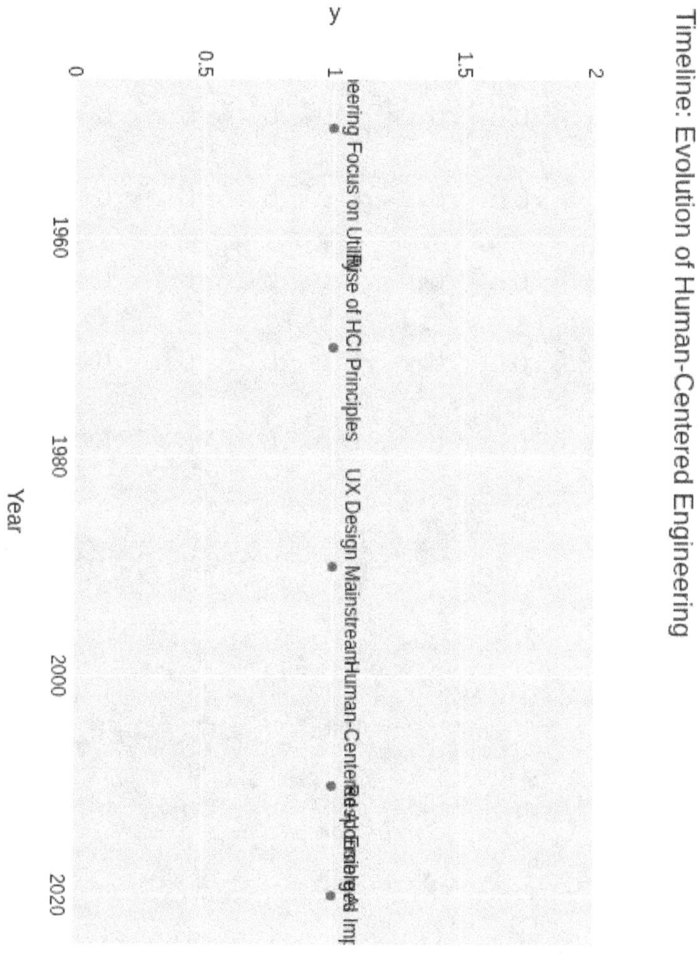
Timeline: Evolution of Human-Centered Engineering
y
0
0.5
1
1.5
2
1960
1980
2000
2020
Year
UX Design Mainstream

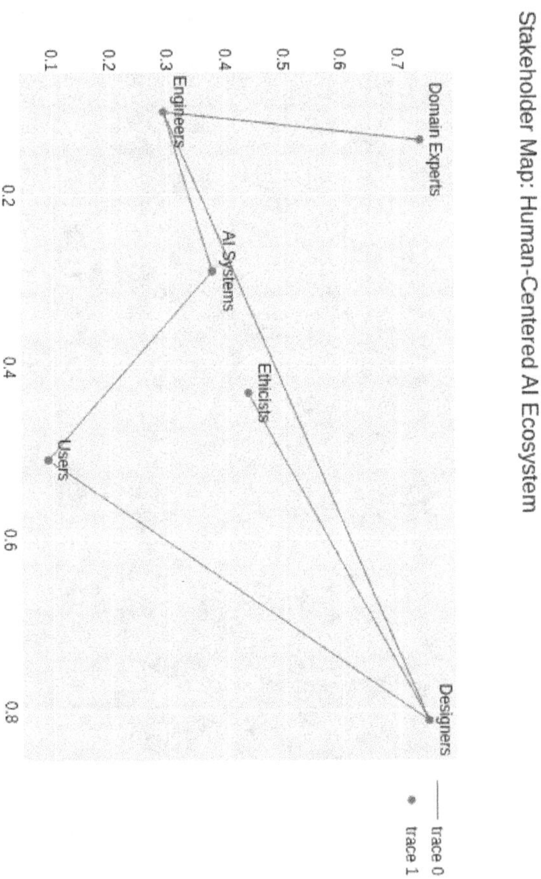

Stakeholder Map: Human-Centered AI Ecosystem
Domain Experts
Engineers
AI Systems
Ethicists
Users
Designers
0.1
0.2
0.3
0.4
0.5
0.6
0.7
0.2
0.4
0.6
0.8
trace 0
trace 1

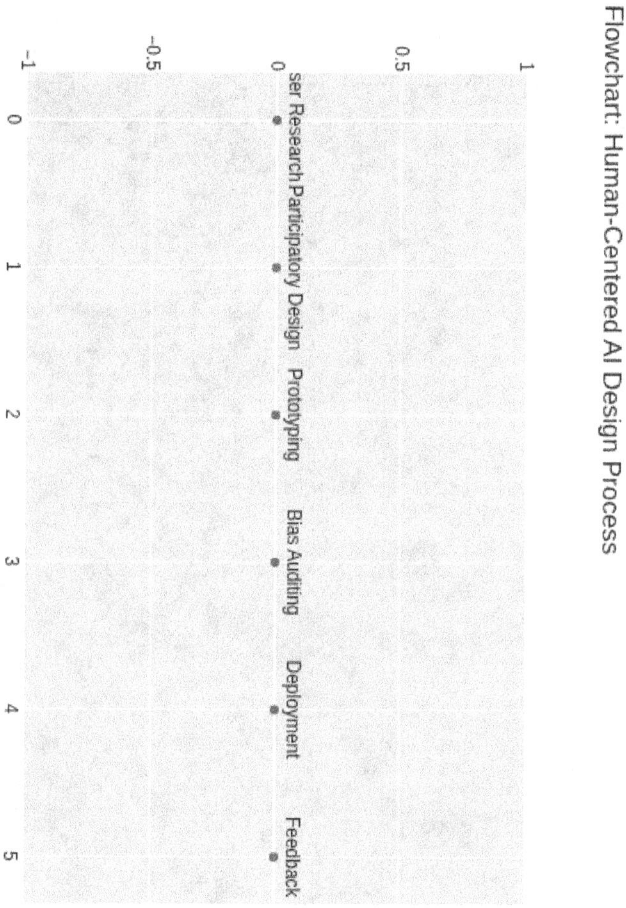
Flowchart: Human-Centered AI Design Process
ser Research
Participatory Design
Prototyping
Bias Auditing
Deployment
Feedback
1
0.5
0
−0.5
−1
0
1
2
3
4
5

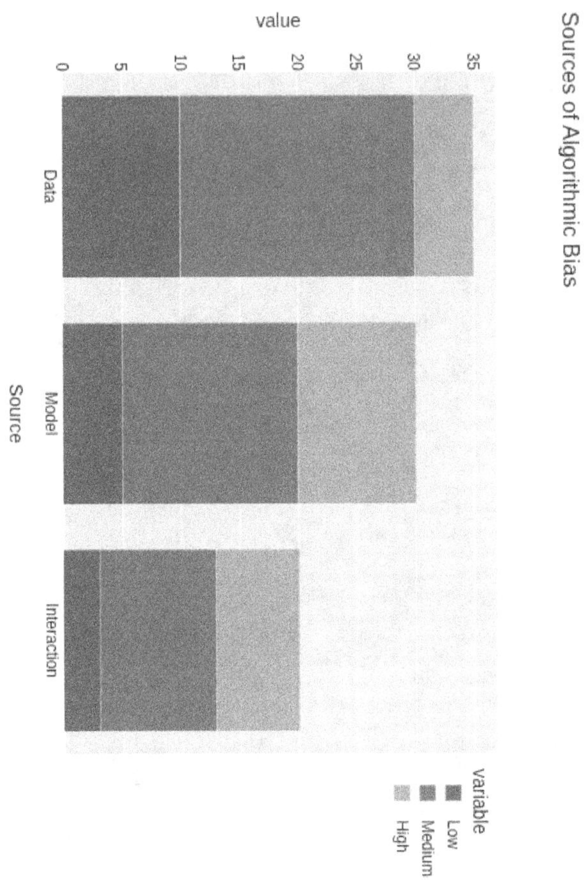
Sources of Algorithmic Bias
value
0
5
10
15
20
25
30
35
Data
Model
Interaction
Source
variable
Low
Medium
High

Bias Mitigation Strategies Across AI Lifecycle

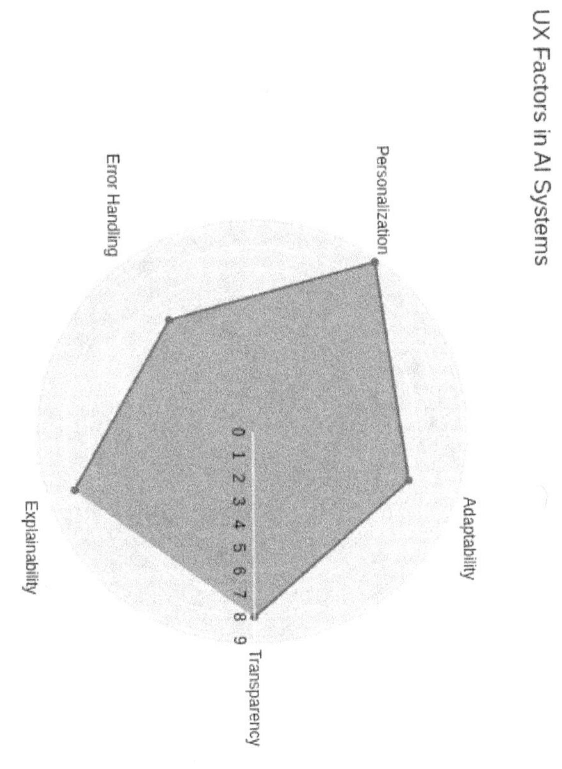
UX Factors in AI Systems
Personalization
Adaptability
Transparency
Explainability
Error Handling
0 1 2 3 4 5 6 7 8 9

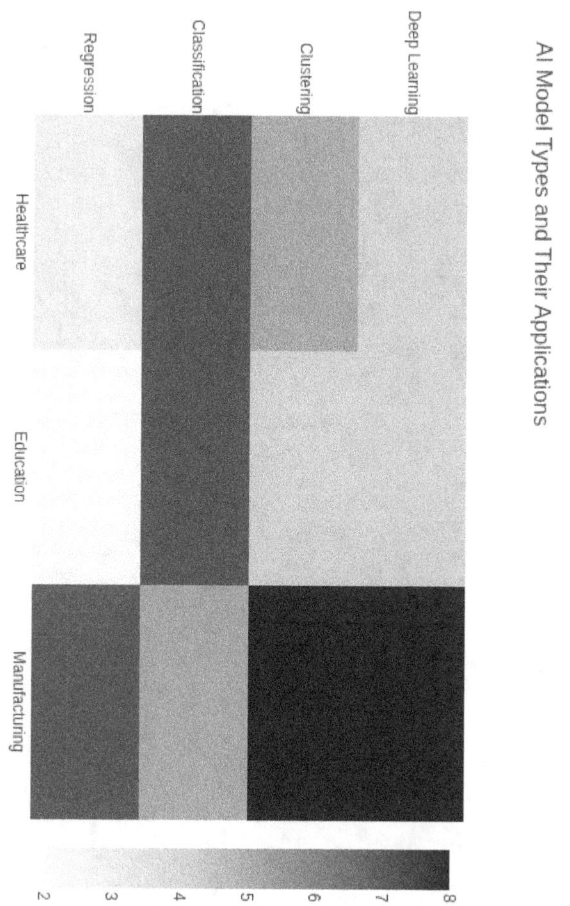
AI Model Types and Their Applications
Deep Learning
Clustering
Classification
Regression
Healthcare
Education
Manufacturing
2
3
4
5
6
7
8

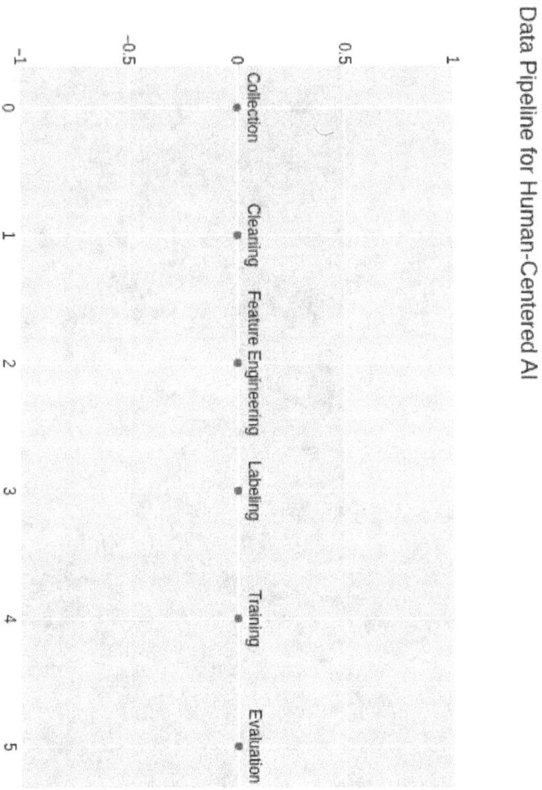
Data Pipeline for Human-Centered AI
Collection
Cleaning
Feature Engineering
Labeling
Training
Evaluation
1
0.5
0
−0.5
−1
0
1
2
3
4
5

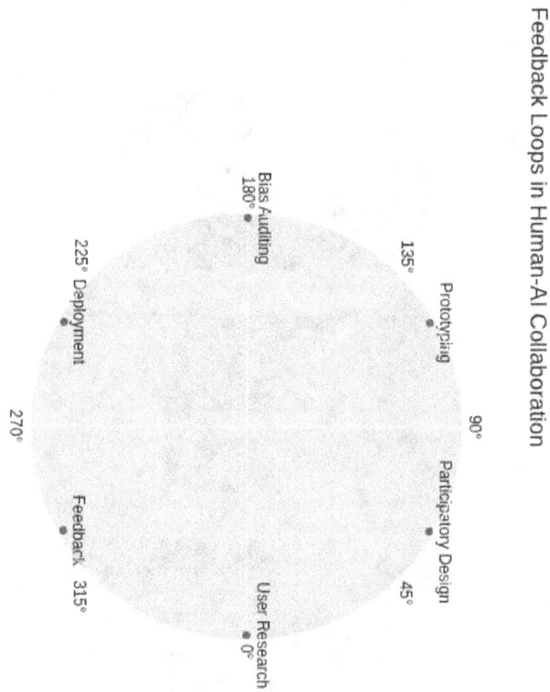
Feedback Loops in Human-AI Collaboration
User Research
0°
Participatory Design
45°
90°
Prototyping
135°
Bias Auditing
180°
Deployment
225°
270°
Feedback
315°

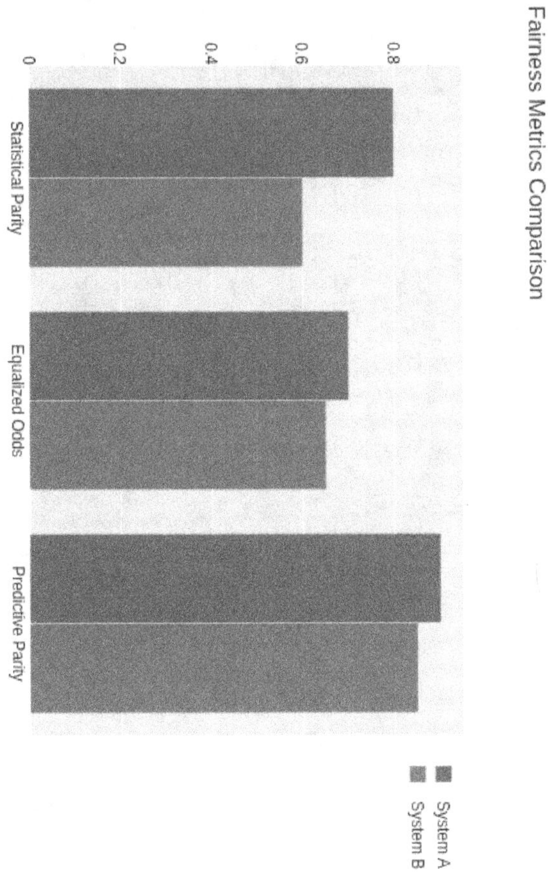
Fairness Metrics Comparison
0
0.2
0.4
0.6
0.8
Statistical Parity
Equalized Odds
Predictive Parity
System A
System B

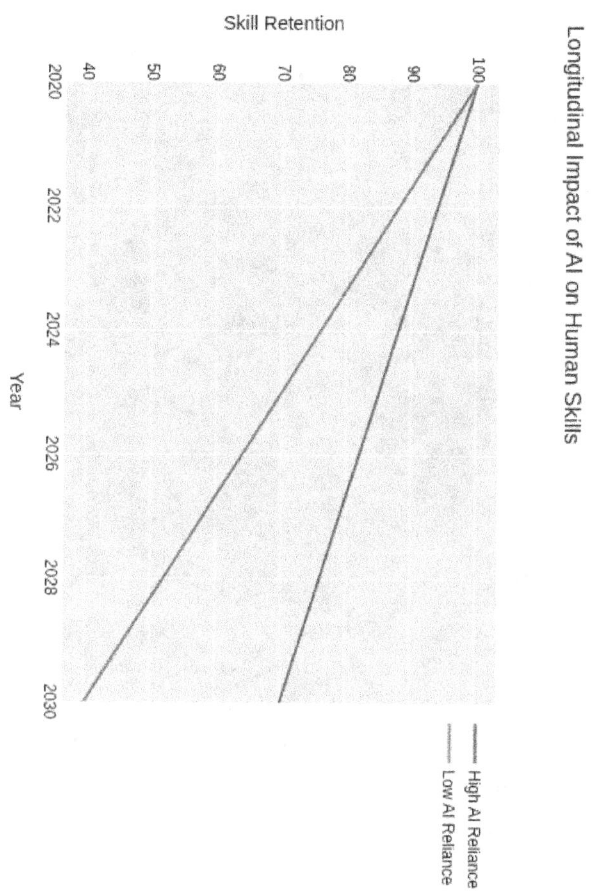
Longitudinal Impact of AI on Human Skills
Skill Retention
100
90
80
70
60
50
40
2020
2022
2024
2026
2028
2030
Year
High AI Reliance
Low AI Reliance

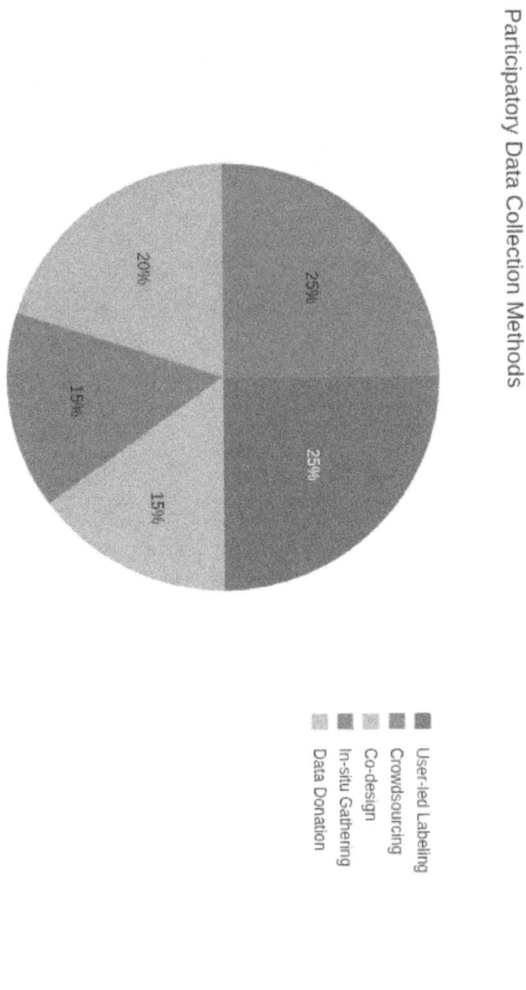
Participatory Data Collection Methods
25%
25%
15%
15%
20%
User-led Labeling
Crowdsourcing
Co-design
In-situ Gathering
Data Donation

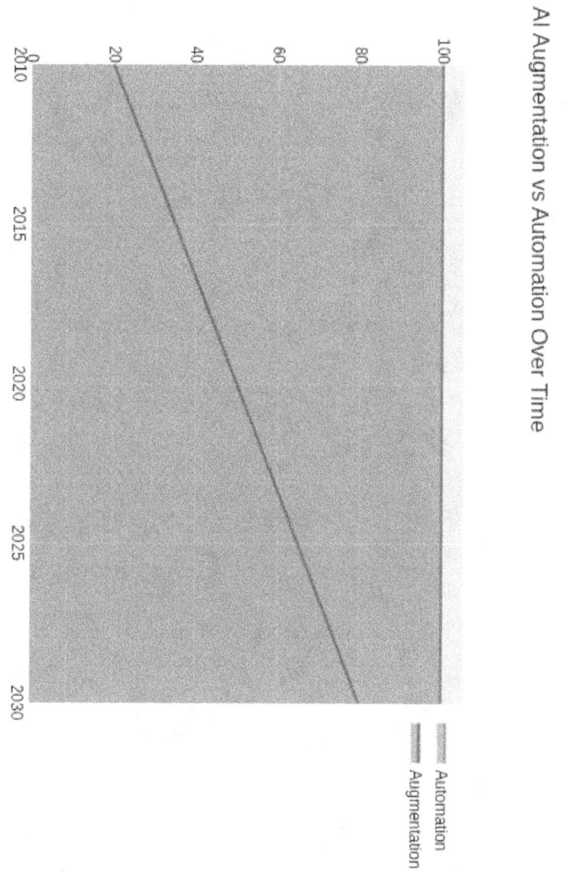
AI Augmentation vs Automation Over Time
100
80
60
40
20
0
2010
2015
2020
2025
2030
Automation
Augmentation

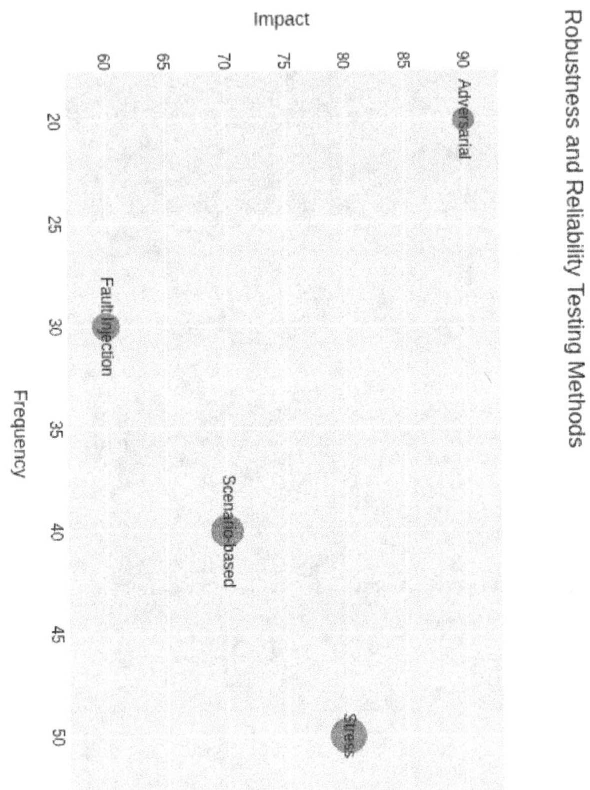
Robustness and Reliability Testing Methods
Impact
Frequency
Adversarial
Stress
Scenario-based
Fault Injection
90
85
80
75
70
65
60
20
25
30
35
40
45
50

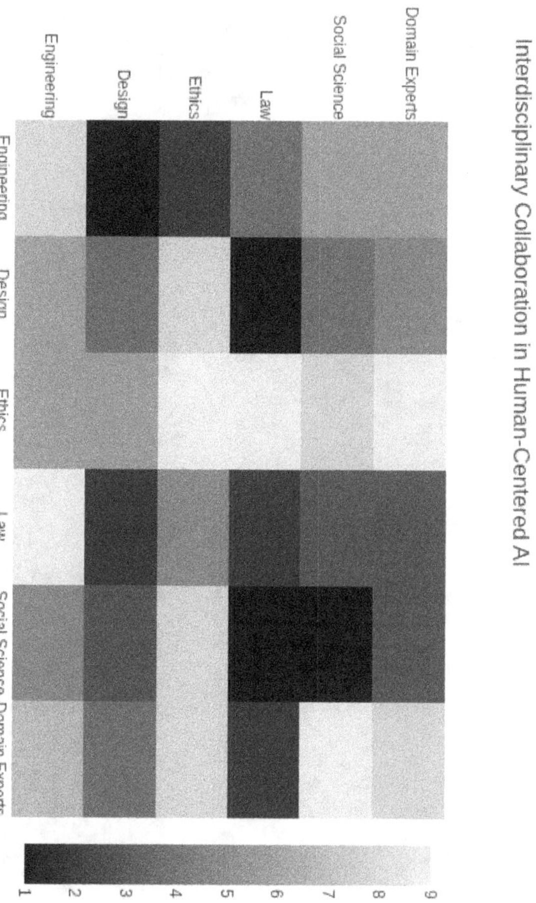
Interdisciplinary Collaboration in Human-Centered AI
Engineering
Design
Ethics
Law
Social Science
Domain Experts
1
2
3
4
5
6
7
8
9

www.ingramcontent.com/pod-product-compliance
Lightning Source LLC
LaVergne TN
LVHW052015200926
848481LV00005B/1497